Time, Emergences and Communications

Engineering, Energy and Architecture Set

coordinated by
Lazaros E. Mavromatidis

Volume 4

Time, Emergences and Communications

Bernard Dugué

ISTE

WILEY

First published 2018 in Great Britain and the United States by ISTE Ltd and John Wiley & Sons, Inc.

ISTE Ltd
27-37 St George's Road
London SW19 4EU
UK

www.iste.co.uk

John Wiley & Sons, Inc.
111 River Street
Hoboken, NJ 07030
USA

www.wiley.com

FSC
www.fsc.org
MIX
Paper from responsible sources
FSC® C013604

Library of Congress Control Number: 2018932804

British Library Cataloguing-in-Publication Data
A CIP record for this book is available from the British Library
ISBN 978-1-78630-250-2

Contents

Introduction

Presentation of the work

It is common for an author to introduce the main lines of his work by placing it in the context of a specific disciplinary field. If the author suspects that his work will challenge established conventions, it is sometimes preferable for him to announce his intentions beforehand, and to develop a series of hypotheses and conjectures. Such is the case of this book entitled "*Time, Emergences and Communications*", which does not fit into any of the two conventional categories, neither that of works which develop analytical studies nor that of synthesis essays. This book anticipates knowledge by outlining a scientific philosophy destined to interpret things and entities that have come into being or which are in a development process, while formulating general and bold hypotheses about those things which have emerged, those which have submerged, as well as those concerning the root of that which still escapes us: time.

We will endeavor to understand emergence in various situations where material components interact by coordinating their actions in order to "constitute systems" with emerging properties (or functions) accessible to experimental investigation or social analysis where man is involved. I will strive to show that communications play a key role in these processes and for that, a strategy will be implemented. If communications are so important, then it must be shown that they are an essential property of matter, which justifies the detailed analyses of the quantum world developed in the first five chapters. If we want to understand the functioning of complex systems such as living cells or animals, we cannot circumvent the study of the strange property of entanglement as well as the interpretation of chemical bonds. The strategy is thus clear. It consists of explaining what quantum matter really is, and then moving on to realities made up of very large numbers of elementary components.

After five chapters devoted to the interpretation of quantum physics, Chapters 6 and 7 will attempt to explain how emergences occur in large sets of elements by focusing on the phases of matter. Contemporary physics is rich in detail, making it possible to reveal the causes of emergence and to bring out this little-known aspect of contemporary science: quantum and material communications whose role appears via the theories of condensed matter as well as quantum field theories. Emergences and communications, these two notions harmonize with each other and work with a third term that still remains enigmatic: time (with its arrow, its polarities and its orientations). Chapters 9 and 10 will explore the question of time, after a brief discussion about computing in Chapter 8. Chapter 11 will offer a general view of the different emergence categories by describing two orders of reality constituted from quantum microphysics (the infra-physical): the mesoscopic level (living cell, viruses, etc.) and the macroscopic level, which corresponds to our daily universe, populated by living species and in which human societies flourish.

Summing up, Chapters 1 to 5 offer an expanded study on communication in quantum matter. Then, Chapters 6 to 8 focus on emergence and Chapters 9 and 10 deal with the question of time. A complete study on the topic of emergence is presented in Chapter 11. Chapter 12 concludes this study devoted to the triplet: *communications, emergence and time*.

The aim of this book is to explain how natural information is exchanged through the fields of communication and is involved in every emergence found in matter, life, societies and the cosmos. If information is a means of producing emergences, it is also at the root of time. This book will introduce unprecedented research. It resumes the quest begun in my previous essay on information, while taking up and completing the fundamental thesis that was delineated therein, namely the duality of physical processes, involving arrangements and communications. In particular, this study will introduce an interpretation of quantum mechanics which, to my knowledge, is unprecedented and which starkly contrasts with most of the classical writings about this enigmatic science. Thermodynamics will also be approached in an unconventional way, in terms of noise and information, which will lead us to examine the roots of time.

A scientific philosophy

Therefore, the essay that you have in your hands is a scientific philosophy, quite a different discipline from the philosophy of science and the history of science. These two fields, sometimes included under the notion of epistemology, envision science and the results it yields, as an object of study. They share a common interest with the history of philosophy, which, like an archeology of meaning and concept, studies transcendental thought as it spread over the centuries. What is more, the philosophy of science studies its object by observing it, analyzing it, but without modifying its object or its formalisms and theories. Contrary to this, the scientific philosophy that I propose is based on scientific results that will be interpreted and selected to serve the purpose of understanding nature and its emergences. Scientific philosophy does not hesitate to upset scientific theories in order to deliver the secrets of nature, and particularly, this strange material that quantum physics studies.

The elaboration of knowledge outlined in this essay concerns each of the central notions mentioned so far: Time, Emergence and Communication. This naturally suggests the idea of three anticipated upheavals: (1) the quantum revolution of communication, (2) the semantic revolution of emergence and (3) the biographical and cosmological revolution of time. All this together with the hypothesis of a Trinitarian conception of time that might lead to a new enigma, even a mystery whose understanding must be sought.

We must understand this book as an attempt to bring together distant disciplines by developing a philosophy capable of building bridges between some central notions, particularly that of communication. In this sense, our approach assumes the legacy of the 20th Century systemic currents gathered around the concept of self-organization and whose ambition was to find the common ground that would help harmonize statistical physics with the socio-political spheres, passing by biology. In the end, I hope to have convinced the reader of the preponderance of communications in the genesis of the emerging worlds, from quantum to the cosmos and the logos, without forgetting about consciousness and the emergence of mental processes.

One last precision, though. Scientific philosophy as I perceive it justifies a particular use of bibliographic references that are sometimes distant from more academic works. This essay has been written in a "Bergsonian" spirit, not only guided by a certain vision of nature, but also using some remarkable results of contemporary science, especially those related to the fields of physics and biology. Finally, the transversal intention of the work may give the impression of a dispersion of arguments; however, a careful reading of the text will show that the ideas developed in it are not arbitrary, but follow a consistent conceptual coherence.

Time, emergences and communications is not a concluding book, but an essay that opens new paths and avenues, both in scientific research and in philosophical thought. In a certain way, it is a manner of restarting the construction of the real. This possibility was proposed by Popper in his preface to *The Logic of Scientific Discovery*, published in German in 1934: "Indeed, it has now become a recurrent question in philosophical circles whether philosophy will ever go far as to pose a genuine problem. Nevertheless, there are still some who do believe that philosophy can pose genuine problems about things, and who therefore still hope to get these problems discussed, and to have done with those depressing monologues which now pass for philosophical discussions. And if by any chance they find themselves unable to accept any of the existing creeds, all they can do is to begin afresh from the beginning" [POP 73].

I hope to have configured three authentic problems, that of quantum matter, then the question of emergences and, finally, the enigma of time which is here conceived as Trinitarian. I propose a new way of understanding things refusing the ontological limits of modern atomism and envisioning information as a basic element of the universe, in that it circulates and is ordered according to different modalities, rules and forms. Then, it is a return to Heraclitus that is put forward, together with the famous formula commented by Hadot about a "nature who loves to veil itself" [HAD 04], except that the aforementioned return is not understood as a regression. Unveiling and veiling correspond to two orders of reality that have been succinctly outlined: emerging phenomena and the submerged structures which are responsible for organization and for emerging forms.

Acknowledgments

I would like to kindly thank Lazaros Mavromatidis for his suggestion to study emergence processes and for having offered me the chance to publish this book. I would also like to thank Anne Pavan for the careful proofreading of the text as well as the corrections suggested. Finally, I wish to thank Julieta Schroeder for her help in the production of this book.

1

Quantum Physics and Information

1.1. Orthodox introduction to quantum physics

"Nobody understands quantum mechanics!" This brilliant catchphrase uttered by Feynman is still true. The dozens of books devoted to this enigmatic science have not yet provided an explanation regarding the quantum world, but the descriptions of these strange experiences as well as these esoteric theories are nonetheless pleasant to read, provided that we are equipped with a minimum of mathematical baggage to travel without much difficulty in a universe that refuses to reveal its secrets. That being said, classical mechanics are hardly easier to understand when they are spiced up with abstract notions such as the configuration space, the Lagrangian or the Poisson bracket [SUS 15]. Electromagnetism is even less accessible. If the world of quantum mechanics is difficult to understand, it naturally follows that introducing this science to a wider audience constitutes a real challenge for physicists, especially considering that physicists do not fully grasp the strangeness of the quantum world. How is it possible to teach quantum mechanics if nobody understands it?

The first book that explained the physical and mathematical principles of quantum mechanics was published by Dirac in 1930. Reading this text, we are struck by the clear and elegant presentation of early quantum mechanics, a few years old at that time. Historians of science have suggested that Dirac might have been affected by a certain kind of autism, which might explain his ability to capture the principles of these new mechanics early on and to expose them with undeniable elegance, together with a very platonic aesthetic concern. Dirac exposed the basic quantum mechanics, including the relativistic equation of the electron. In 1930, the great enigma of quantum entanglement was not known and specialists were satisfied with the orthodox interpretation acquired after bitter controversy during the Solvay Congress of 1927.

A presentation of quantum mechanics should also include the so-called orthodox interpretation. Quantum mechanics are integrated by two blocks. In the first place, the preparation of the experiment described in the form of a superposition of states, each described by a complex function, ψ_1, ψ_2, ψ_N, etc. The notation system uses the column vector (bra) or the line vector (ket), symbolized by a | bar, respectively followed or preceded by > or <; each state corresponds to an observable[1]. Let us imagine an experiment with two observable colors, red and blue. The system is described by an addition of the two vectors; each vector is assigned a coefficient whose square corresponds to the probability that the observable associated with its state may actually appear in the experiment. One of the requirements is that the sum of the squares equals one.

$$\text{State vector} = \sqrt{1/3}\ |\ \psi\text{-red} > + \sqrt{2/3}\ |\ \psi\text{-blue} >$$

– This state vector indicates that there is a one in three chance for red to appear and two chances out of three chances for blue to appear. When there is no observation, the wave function is supposed to obey a deterministic equation established by Schrödinger. Evolution is continuous in time and deterministic. When a measurement is made, there is a collapse or a reduction in the state vector. The measurement is discontinuous and indeterministic because we cannot know in advance if it is red or blue that will pop out of the “quantum roulette”.

– The orthodox interpretation states two things:

1) First, the wave function (or vector) describing the system does not correspond to a physical reality of the world (be it objective or not). It is merely a mathematical tool for calculating what is

1 In physics, an *observable* (noun) is a dynamic variable that can be measured.

observable and the probability associated with each measure. In my example, the wave function indicates that the moment when the experiment is performed, blue and red will be able to manifest with one-third and two-thirds probabilities. However, the wave function does not assign any property to the system, which is neither blue, nor red, nor purple. In fact, the wave function does not have a "color" in the space of quantum vectors. In other words, we know nothing about the system in terms of the characters or phenomena present in our world.

2) Second, once the experiment is launched and the measurement made, thanks to the detectors of our world, we know perfectly well whether it is the red or the blue that will be expressed. The orthodox interpretation separates the description of the system when nobody is trying to make an observation and what occurs if an observation is made and the wave function is reduced (in order to express red or blue). According to Bohr, the quantum experiment might require two descriptions: an independent one (purely quantum and made of wave functions with no real counterpart) and a classical one, including the laboratory and its devices. The whole enigma can be reduced to what happens when we move from the "unreal" quantum stage "$\sqrt{1} / 3 \mid \psi\text{-red}> + \sqrt{2} / 3 \mid \psi\text{-blue}>$" to the classic "blue" or "red" stage. Besides, the orthodox interpretation postulates that we know nothing about a system until we have measured anything. And therefore the most complete description of the system, that of the wave functions, is not expected to belong to the field of the natural knowledge of things. Only "blue" or "red" signals belong to our material universe, which can be known by performing an experiment.

While the orthodox interpretation has been accepted by the majority of physicists, other interpretations have been offered. For example, many-worlds theory assumes that if we observe blue during a reduction of the wave function, then the universe splits in two, with a double in which red is observed, except that we do not have the means to go and check what happens in the split universe and verify if red has appeared there. It is nonetheless possible to conduct such a precise experiment that we are able to observe blue and red superimposed in a system for an infinitesimal period of time. This is the decoherence experience that was achieved in the 1990s and recently won a Nobel Prize for Serge Haroche and David Wineland. Decoherence does not change a word to the orthodox interpretation, but it does question the link between our classical world and the quantum world. In fact, it arouses bold speculations about the measurement process that could take place much like a natural selection of observables, from which derives the concept of "quantum Darwinism" (that we will explore in Chapter 4). Scientists who want to know what occurs during the measurement instance are similar to the spectators of a prestidigitation act, wishing to go behind the scenes in order to understand how the magic trick is carried out. They want to go beyond the horizon and lift the sails. If primordial matter is a quantum gum, then nature unfolds as if it were chewing multicolored gum. Then, the experiment consists of extracting red or blue bubble gums that stick to our extended matter and color it. Why and how? This enigma is not solved yet.

There is a more classic image that can help us to represent this strange wave function and its reduction at the moment when it is measured. Let us imagine a football game. Zidane is in possession of the ball. He can choose between four possibilities: either to keep it, to pass it to a winger, to pass it to a forward or to a fullback. The match is filmed but it suddenly stops. The wave function that determines the experiment is defined like this, assuming that each possibility is equiprobable.

> ψ-Zidane = ½ | to keep it > + ½ | pass it to winger > +½ | forward > +½ | full back >

In order to find out how this continues, we have to resume the film. A few seconds after the retransmission was interrupted, we then see Zidane's choice. The operation for reducing the wave function took place in Zidane's brain. However, the picture is misleading. The description of the system by ψ-Zidane deals with classical situations, not with quantum waves. These are classic probabilities we are referring to, as in a game of dice. The terrain where wave reduction takes place is a classic, macroscopic scenario, whereas in the quantum world, if this "terrain" is a reality, it is partly out of our classical scene. Besides, the evolution of ψ-Zidane's brain wave function depends on the information collected by Zidane. If the forward player is marked by an opposing defender, but the opposing winger progressively stops marking, the wave function necessarily changes; the probability of passing the ball forward decreases while that of passing the ball to the winger increases. In quantum world, such an eventuality is inconceivable because as long as the system does not exchange information, the evolution of the wave function remains independent from what is going on in our classical world and obeys Schrödinger's equation.

The orthodox interpretation has been discussed in thousands of articles. It suggests a limitation of natural knowledge, that is, it draws a horizon of knowledge, a concept cleverly introduced by Gonseth. Experimenting with decoherence has slightly pushed this horizon, as occurred with the recent works of Humphrey Maris, who detected phenomena that may be interpreted as a fission of the wave function under extreme conditions, those of superfluid helium (see Chapter 2, section 2.4). Thus, quantum description in terms of complex waves corresponds to an element of the physical world under exceptional conditions. In a totally different context, quantum entanglement has equally pushed the horizon of knowledge. In a device containing two intertwined "elements", a measurement on one of them makes it possible to obtain information about the second one. In other words, global information is contained in the part (see Chapter 3).

As a result, these considerations open a dialogue with nature: in fact, what is matter from the quantum point of view? Having elaborated this question in other papers, I suggest some elements of response by refocusing on the notion of information, which will be a common thread in this work. Primordial quantum matter "produces" or "contains" information that can be communicated to the experimenter by means of a technological interface. Therefore, following the previous example, the information content is provided by a wave function that looks like this: $\sqrt{1} / 3 \mid \psi\text{-red}> + \sqrt{2} / 3 \mid \psi\text{-blue}>$. The information communicated corresponds to one of the observables, red or blue. In other words, it is quantum gum that sticks and is introduced into the technological device. It is in a similar fashion that hidden impressionist matter projects a few tiny drops of paint, if we look close enough. By nature, quantum physics is a dynamic. However, it is a dynamic of communications, not of mechanics with forces and is arranged in a geometrical extent, like that of Newton or Lagrange.

1.2. Quantum states or how nature communicates with physicists

Let us start with a general idea that is applicable to every physical experience; two notions are always present: state and measurement. In a sense, a state is like a "dynamic and kinematic inventory". With variables and equations that determine dynamics and kinematics, that is, the arrangement of forces and the movement of particles. A measurement is a number obtained during observation. It may be a speed, a position, an angle, a mass, a trajectory, a translational or torsion force, kinetic momentum, a particle detected with a spin, etc. The analysis of the "epistemological work" around the notions of state and measurement makes it possible to highlight the irreducible difference between classical physics and quantum dynamics.

Firstly, the definition of a quantum system or a quantum state with wave functions is completely different from the descriptions used in mechanics. Second, we have to be aware that while in classical mechanics the notions used for describing the state of a system are also implemented for describing observations (speed, angles and positions); in quantum mechanics, this is no longer the case. I will now illustrate with the example of a function defined as the Lagrangian in classical mechanics. To calculate it, we have to subtract the kinetic energy from the potential energy and its formula can be summarized as $L = L(p, q)$. In this formula, it is sufficient to know the potential and six parameters, three for p, the position, and three for q, the impulse, the latter being calculated from m mass and dp/dt speed. In a way, the Lagrangian sums up the state of the system, as it evolves along a trajectory by obeying the principle of least action. It is clear that the parameters used for describing the state are also the parameters used in experimental measurement (mass, speed and position). If we now consider a quantum state described by the formula:

$$\Psi = \sqrt{1/3}\ |\ \psi\text{-red} > + \sqrt{2/3}\ |\ \psi\text{-blue} >.$$

We find that $|\ \psi\text{-red}>$ and $|\ \psi\text{-blue}>$ are mathematical formulas that do not describe real things in the experience. On the contrary, the observables, red and blue, are "real". In classical mechanics, the evolution of the system is calculated from the Lagrangian, whereas in quantum mechanics, it is the Ψ wave function that evolves by "obeying" the Schrödinger equation which is applicable as long as the system is not observed (that is, disturbed). This is what makes quantum physics strange, since the evolution of the system concerns something that is not observable, and what is more something which does not correspond to a physical reality. When observation is carried out, the classical rules of determinism are no longer valid.

The link between the two domains is neither intuitive nor trivial. The world of observations is deduced from that of the states by means of rather esoteric mathematical operations. Man and higher animals are naturally "conceived" for perceiving macroscopic classical things, force, figures, position, temperature, but not quantum phenomena. Scientists have had to develop an arsenal of mathematical and technological tools to describe and observe quantum systems. In the quantum experiment, every observation modifies the state of the observed system. On the contrary, in classical physics, the observed system is considered independent from the observer. The modification of the quantum state is interpreted as a two-way communication. In other words, the exchange of information is reciprocal up to such an extent that when the matter communicates information to us, we also communicate information to the matter, which has in this way been a recipient of our influence. Let us recall that the notion of influence over a system or a thing is one criterion used for characterizing and for defining the ambiguous and ambivalent notion of information in the physical sense. How do we extract this information? Let us take the example of the system described by:

$$\Psi = \sqrt{1/3}\ |\ \psi\text{-red} > + \sqrt{2/3}\ |\ \psi\text{-blue} >.$$

To go from the state vector to the observable, we apply an operator to the vectors, which are then transformed. Given a C operator (which will be called "color operator"), this has a very special property, found in all quantum physics operators, namely that it is Hermitian. This means that if it acts upon a state vector, the result is a multiplication of this vector by a number that is real and not complex. And this is convenient because a real number may correspond to an experimental measurement. Let us write the application of an operator to the proper vector corresponding to red:

$$C \mid \psi\text{-red} > = \text{Red} \mid \psi\text{-red} >.$$

We can do the same for the other proper vector:

$$C \mid \psi\text{-blue} > = \text{Blue} \mid \psi\text{-blue} >.$$

Color operator C makes it possible to extract two observable properties from the system described by Ψ, which are red and blue. This operation should be interpreted as taken from the formal viewpoint of objective information from a world which is not objective itself, that of state vectors Ψ. What is so particular about a Hermitian operator? In linear algebra, a Hermitian operator is defined as a self-adjoint endomorphism. This means that the C operator and its designated assistant indicated by a *dagger* (or C^+, read C cross) match, which gives the formula $C = C^{dagger}$. Quantum operators are often matrices. Obtaining an assistant requires two operations, transposition and conjugation. Transposition refers to tilting the matrix around its diagonal. It is as if we were turning a metal square using its diagonal axis. In a matrix, the terms are noted with two indices which designate the row and the column. Transposition leads to reversing the indices. For example, the element placed in the third row, second column moves towards the second row, third column: thus the transposed c_{32} is equal to c_{23}. On the contrary, conjugation refers to reversing the sign of the imaginary part of a complex number. For instance, if $c = 7 + 3i$, then its conjugate is written with an asterisk, in such a way that c^* equals $7 - 3i$. Thus, we have the formula: $c_{32}^* = c23$, and more generally, $c_{ab}^* = c_{ba}$. The elements of the diagonal have no imaginary part and are real.

Given a square matrix with $c_{21} = 1 + i$; if we want the adjoint matrix to be identical to the initial matrix, then $c_{12} = 1 - i$. The other two coefficients are in the diagonal and must be real numbers, as for example $c_{11} = 3$ $c_{22} = 4$. This results in the following matrix:

$$\begin{matrix} 3 & 1-i \\ 1+i & 4. \end{matrix}$$

Therefore, let us now imagine the square matrix as a piece of metal whose shape is square and where each term is represented with a surface corresponding to the whole part. Now, this part is deformed with a bump (or a hollow) whose height (or depth) corresponds to the complex part. The sign of the imaginary number (positive or negative) determines whether it is a hump or a hollow. In our example, the hump of coefficient $c_{21} = 1 + i$ mirrors the hollow, which is associated with $c_{12} = 1 - i$. Hence, if we turn around the metal square choosing its diagonal as rotation axis, then what is a bump with a certain height becomes a hollow with a certain equivalent depth, and as a conclusion, rotation does not change the shape of the metal square.

Are Hermitian matrices simple calculation tools or do they have anything to say in regard to this rather enigmatic matter that we can conceive as vibrating quantum? If we assume that quantum states are in the image of structured vibrating gums, then the shape of the operator describes a process capable of making a reversal operation look like a gum element from an interior place towards another crossing point in the quantum interface and an extension in our macroscopic extended universe. Quantum gum communicates by spreading over the extended world and, conversely, with the symmetry of reversal we may think that the information of our world is transmitted to matter. If this matter could "observe" what is going on, it would see the same observable that we see at the laboratory. As if it looked at us through a mirror.

The Hermitian character of the operator reveals a symmetric property that characterizes "quantum nature". By symmetry, we mean an operation makes an object move without changing its shape. This symmetric reversal might explain why we perceive things with our sight. Therefore, the thesis of the visible world that fixes itself on the retina after passing through the lens and that the brain recomposes is a fable. Sensitive vision can be explained through quantum physics, to which the principles of extended macroscopic order are added. We will note a coincidence: the geometric tensors of general relativity are 4×4 matrices but with real terms; these can be reversed, taking the diagonal as axis of rotation without them changing their shape.

1.3. Particles, information, evolution

Quantum dynamics can be described by means of an epistemological triplet: state vectors, operators and observables. State vectors define overlapping "quantum states" when these are not observed. In his treatise on quantum mechanics, Feynman has stated that any state in the world could be described as an overlapping of basic states and we wonder which are these basic states that exist in innumerable situations, as for example in the spinning of an electron or a proton. These examples bring us back to natural order, that is, to physical order. If the three elements of the triplet wave, operator and observable triplet are no more than mathematical forms, then the experience of the world and the quantum experience lie upon a natural physical substratum that we call matter and which manifests itself in an extended field of expression. If we assume that the states of the world constitute basic information, then the particles act as the medium for expressing such information. This conception brings us back to Aristotelian hylemorphism with a form–matter duality which, in the quantum world, becomes the principle of a duality between quantum states and particles, understanding that a particle (in terms of mass or energy) belongs to the material pole. Quantum neohylemorphism as studied in my thesis [DUG 96] will be exposed in Chapter 2.

Hence, I interpreted fundamental quantum matter as an overlapping of information arranged in a certain order. This information is formally extracted by operators and can be found in experiments in the form of observables. We can reasonably think that the "material particle" is the physical element that performs this extraction, which updates information in our world. In this way, basic states are somehow the DNA equivalent to the primordial material world. The operator works as a quantum code expression system, in the same way as the genetic code is expressed in RNAs. Observables would then correspond to proteins translated from RNA. Thus, quantum dynamics is the science that describes the expression and communication of information by means of a matter that takes part in experimentation as a set of particles with defined features.

Quantum physics teaches us that states evolve over time. While the triplet "vector, operator and observable" is complete for describing quantum measurement, time evolution concerns the state vector. This evolution is shown by the Schrödinger equation ($H\ \Psi = E\ \Psi$). In classical mechanics, energy is the state variable as this is preserved through time (see Noether's theorem). The energy makes it possible to calculate the classical Hamiltonian. In quantum mechanics, the Hamiltonian is not an amount but an H operator whose proper value determines the E energy that the system "consumes" when it is observed. Let us nonetheless recall that in Heisenberg and Jordan's *Matrix mechanics*, it is the observable that evolves over time (see Chapter 2), which finally changes nothing because the two versions are equivalent.

The evolution operator noted as U is unitary. The result of applying a U operator and then its auxiliary leads to the same state. In other words, the multiplication of the two operators is the identity operator. $U \times U^{dagger} = I$. With one important consequence: the "logical" relation between two states is preserved by time [SUS 16]. This property presents a similarity with a Liouville theorem that states the conservation of the number of configurations over time for an irreversible dynamic system or even another Liouville theorem on the conservation of the space volume between phases. As a unitary operator, Liouville's theorem and many other things are connected to the issue of information conservation. As soon as we approach the question of information, the links between quantum mechanics and other physics become more precise and this is also the case of the AdS/CFT[2] correspondence in quantum gravity.

---

2 Anti de-Sitter/Conformal Field Theory (AdS/CFT).

We will stick to the idea of order in how basic quantum states are arranged, that is, an information order of "fundamental matter". This is what makes it possible to propose a universal principle. In a classical system, the state variable preserved in time is energy, provided that this system does not exchange energy with the environment. *In a quantum system, it is information that is preserved.* Then, the evolution operator has a dual role. It determines the evolution of quantum states while it preserves the order of information and it makes it possible to calculate the energy associated with each state. In other words, an expressed state manifests itself as information extracted by an operator and also from a different angle, with energy necessary for its expression in the material world. This is what makes it possible to envision equivalence between information and energy.

1.4. Interpretation of Pauli's exclusion principle

One of the strangest properties of the quantum world is explained by Pauli's exclusion principle. Two particles cannot simultaneously occupy the same state characterized by quantum numbers and a level of energy. This principle concerns the particles having a half-integer spin, which is the case for the electron, the proton and the neutron, particles that have a spin equal to ½. These particles are called fermions. Thus, a set of indistinguishable particles, placed under conditions where the quantum effects are effective, are placed on the energy levels foreseen by the description of the system and the calculations. This starts with the lowest energy states. This principle elegantly explains the properties of atoms in the context of quantum chemistry (see Chapters 2 and 5). The electrons only have to "place" themselves on each orbital with their energy and a shape is calculated using three quantum numbers.

Particles placed in quantum conditions are distributed obeying a statistical law, Fermi–Dirac statistics, when it comes to fermions. Whole-spin particles such as photons obey another statistic, Bose–Einstein statistics. That is why they are called bosons; these "particles" can massively occupy the same state, in this case the one with the lowest energy. Fermi–Dirac statistics, formulated in a relativistic context, are often introduced as the successful expression of Pauli's exclusion principle. Sometimes the exclusion principle is not considered a fundamental element of quantum physics, although it may have intrigued more than one scientist, including Feynman who speculated in his treatise on physics about what might happen at the level of the atom in the absence of this principle. Although this speculation may possess heuristic value, it does not reveal the secrets of fermions, regardless of the fact that they exist as objects.

In 1929, Pauli's principle experienced a singular episode when Dirac formulated the relativistic wave equation for ½ spin particles, especially for the electron. Dirac's results astonished the specialists of the quantum world. In fact, Dirac's equation is oversized because only two components of the state vector used are useful for describing the electron. What to do about the other two? The most surprising feature was the emergence of negative energy solutions supposed to describe electrons with a positive electrical charge. At first, Dirac rejected these solutions as non-physical, but when he later coupled the electron with an electromagnetic field, he observed that this electron could switch to a negative energy state. This led him to consider that the universe is composed of positive and negative states. As a result, a strange conception of matter ensued, with quantum states occupied by electrons or not occupied at all, which were defined as vacant or holes.

In the 1930s, negative energy states were a curiosity provisionally solved by applying Pauli's principle. Dirac assumed that these states were all occupied and that, by virtue of the principle of exclusion, positive-energy electrons, those of our real world, could not topple over into the already filled negative holes. These explanations are now over and Dirac's Sea Hypothesis no longer appears in modern physics textbooks for fear of making students more confused. Nevertheless, some of Dirac's work has opened paths that may not have been thoroughly explored. The combination of Dirac's Sea, Pauli's Principle and the coupling of this sea to the electromagnetic field provide precious knowledge about "fundamental matter".

It is commonly accepted that particles and atoms are the elementary constituents of matter. However, a complete analysis of quantum dynamics makes it possible to envision another possibility for gaining access to fundamental elements. These are the basic quantum states or, more exactly, the elementary information contained in fundamental matter. Let us suppose that quantum information carried by one state is the fundamental element of the real. Explaining the world no longer needs the formulation of Pauli's principle because it is not the electrons that occupy a state but it is the "states" that achieve themselves in the form of an electron, which is the empirical element indirectly accessible to experience. Let us imagine a stock of information as a set of letters, a, b, c, d, and so on. These states are now achieved by electrons and are designated as A, B, C, D, etc. The exclusion principle is then a principle of multiplicity and differentiation. The idea that an electron occupying a state is actually a mental experiment, which does not necessarily have any physical meaning. What is fundamental is the information carried by the state and the energy that enables its realization. The electrical charge means that the information that goes through the range is quantified and responds to precise constraints.

The electron is an interface that enables the communication of quantum information. The existence of the charge reveals the universal rules that govern communication. We may think the same thing about mass. It is the interaction with the electromagnetic field that provides the "hole" with the features of the electron, whose existence cannot be separated from the field. It is as if visible information manifested itself through a universal material medium, in the image of a flat screen with pixels. In this metaphor, the electron would then be like a natural pixel broadcasting quantum information.

1.5. State vectors, science of orientations

Physics textbooks use two definitions for the same notion: the wave function and the state vector, constituted by a linear addition of complex vectors. The notion of vector contains a precise meaning in terms of mathematical formalism because a vector is not a simple collection of numbers, but it possesses an orientation. Nevertheless, this orientation of quantum vectors is not covered by our study, featuring spatial coordinates and time.

The electric field is given by a vector. Spherical symmetry states that when two points M and M' are diametrically opposed to the source of the field located at S at an equal distance, then the electric vector E_M is equal to $-E_{M'}$. If we take a turn traversed by a current of electric charges, a magnetic field is generated and the plane on which is the turn is a plane of symmetry for this field B. Nevertheless, there is no inversion: vector $B_M = B_{M'}$. Physicists say of the electric field that it is polar or even that it is true. Which means that if we transform a physical reality with a reflective symmetry (of the mirror type), then a vector is transformed into its opposite. The same goes for a speed vector. You see yourself in a mirror and you note that as you move towards the mirror, your image moves towards you and it seems to move in the opposite direction. On the contrary, the magnetic field does not obey this inversion by parity, in such a way that it is considered as an axial vector or a pseudovector.

If we consider a complex vector and represent it in an Argand plane, then the parity transformation becomes evident. In the Argand plane, the horizontal axis represents the real part and the vertical axis represents the imaginary part of the complex vector. For example, $2 + 2i$ is represented by a vector whose length is equal to $2.\sqrt{2}$ and whose orientation is 45 degrees upwards. If we perform a reversal, considering the horizontal axis as symmetry, then we obtain the same vector, except that its orientation will be 45 degrees downwards. This vector is actually the conjugate vector of the previous one, $2 - 2i$. These few remarks reveal the important role of orientation in quantum physics, as well as in electrodynamics and magnetism. Matter is not only displayed in a certain manner, but is also oriented. What is more, this orientation is associated with the communications that quantum matter establishes. Orientation in the quantum sense belongs to the physics of elementary communication, and should not be confused with orientation of mass in the context of a physics of arrangements, such as the rotation of a spinning top or a planet in the cosmos.

Orientation is practically universal in the quantum world. Orientation could be found in the electric and magnetic vectors, describing realities at the confines of the classical world. In the universe of "particles", orientation was introduced as spinning, awkwardly defined as intrinsic kinetic momentum. Spinning actually designates the orientation adopted by a fermion. Pauli matrices constitute a basis for spin states which correspond to three axes, x, y and z. When the atomic electrons are represented on the orbitals, they successively occupy the levels in pairs of two, one with a spin whose representation is an arrow pointing downwards and the other with the arrow pointing upwards. I suggest that these opposing spins (in sign and direction) refer to quantum communications in opposite directions, from the manifested world towards fundamental matter (overlapping quantum information, if you will) and from this fundamental matter towards the manifested world. This can be expressed using Nicolas de Cues's terms, *explicatio* and *complicatio*. Complex fundamental matter can be explained, as it unfolds in order to communicate with the world. To be even more graphic, I use the notions of quantum expression and quantum impression. Expression is communication, impression is information. In order to complete the description, I propose the notion of compressed or even zipped information so as to designate the world of basic states formalized by state vectors.

The dynamics of orientations is explicit thanks to three Pauli matrices that enable us to describe the spin operators associated with the internal rotations of the electron. These matrices are designated with the letter σ, and three indices corresponding to orientations in space assigned to them: $\sigma_1 = \sigma_x$; $\sigma_2 = \sigma_y$; $\sigma_3 = \sigma_z$. I suggest imagining a fly moving in front of a window as if observed by an individual at a certain distance, on the ordinate axis (y-axis). The σx matrix corresponds to the flight of the fly on the window along the horizontal axis. The σz matrix corresponds to the same scan on the vertical axis. Finally, σy, located in the axis of the observer, corresponds to the movement of the fly from one side of the window to the other. Thus, the electron appears as a microwave of a communicating interface that can cover the three directions of space. Therefore, orientation is specific to communication as facilitated by "quantum matter", be it in the sense of reception or in the sense of expression.

The importance of orientation in the quantum world is confirmed by the Compton Effect, often introduced from a mechanistic perspective in physics textbooks. An incident photon comes up against an electron associated with an atom that absorbs a portion of the photon energy and is ejected from the nucleus in the form of an ionized electron, while the photon is launched again, but this time losing a small portion of its energy, which produces a difference between the wavelength of the incident photon and that of the reflected photon. The phenomenon is interpreted as inelastic diffusion. This presupposes an epistemological artifice that is not self-evident because, in principle, a photon has no mass and, therefore, it cannot be involved in a shock. The trick is to consider that, given the very short wavelength of the photon (which is gamma radiation), it can be compared to a corpuscle. In physics textbooks, the incident photon is represented on the horizontal axis, it comes upon an atom, the photon goes upwards with a direction measured by a θ angle, whereas the electron goes downwards with a Φ angle. I suggest interpreting this process as an information exchange. Now, the two angles θ and Φ represent the manifestation of this information exchanged during the process. The shock is of little importance. What is remarkable is the importance of the orientation that is emerging in this quantum experience. As we can see, the Compton Effect marks a clear difference between the mechanical world and the quantum world.

Let us imagine a similar configuration in classical mechanics. The atom is represented by a billiard ball. If we push another billiard ball whose impulse vector meets the mass center of the first ball, then an elastic shock occurs and the second ball acquires the energy of the first and starts moving in the direction of the vector. An inelastic shock occurs if we push a ceramic ball towards the billiard ball. The ball may break into two pieces that will get dispersed. This is the situation found in the Compton diffusion, except that the incident ball is not material; it is a radiation of an almost imperceptible wavelength. Everything happens as if the photon broke the atom by ejecting an electron, except that the photon is not material.

Other research projects will help us confirm the importance of orientation in the physical world with the magnetism that is inseparable from orientation, both in the quantum field as well as in the classical world. In his treatise, the importance of orientation had not escaped Dirac, who evoked the orientation of the radiative emission by the electron which is then transmitted to the photon via polarization. Orientation is present not in a spatial but in a formal sense in the canonical transformations that are specific to quantum mechanics, but not analogous to classical mechanics [DIR 32, p. 95]. A canonical transformation concerns the representation of observables. It is as if the physicist changed his formal view angle so as to represent the system. Finally, orientation appears in the great enigma of neutrinos whose (negative) helicity perfectly translates the oriented character of this "particle". Helicity is to the neutrino what polarization is to the photon. The antineutrino has a positive helicity that makes it the mirror image of a neutrino. In the end, we rejoin the fundamental CPT symmetry[3] of the world of charged particles, except that charge conjugation C does not exist since the neutrino is not charged. Conjugation then takes place at the level of helicity.

1.6. Provisional conclusions on quantum mechanics

In conclusion, quantum physics must be considered as an information science and, more precisely, as a science that describes the fundamental rules that quantum matter obeys, whose fundamental and universal property is to communicate with interfaces and with dynamic exchanges between signals that carry energy and information. Quantum formalism with state vectors, Hilbert spaces and operators describe how the observer extracts information from a fundamental, or even primordial, matter of submerged information.

Quantum physics enhances the importance of orientation, which is the character of a substance that leans on its environment in order to communicate. Quantum physics can only be understood by highlighting its features and by making it even stranger than it is. It is necessary to highlight the characters of the quantum world and not to

3 Charge, Parity and Time reversal symmetry.

try to reduce them while we try to find correspondences with classical and mechanical images, as was done by most physicists who wisely misled themselves by avoiding the realities that this science offers us about the essence and behavior of matter. In his physics class, Feynman explicitly recognized the fact that "quantum mechanics is a different kind of representation theory about the world" and that while correspondences may sometimes exist, these are little more than mnemonic tricks, that is, things for helping us associate ideas [FEY 79, p. 205].

On my behalf, correspondences, be them explicit (like those of Bohr) or implicit (with the primitive notions), are no more than semantic means for speaking about a world that is completely strange to us. Using these notions helps us to say things, but in reality, these things do not exist. For instance, the notions of particles and waves, respectively derived from mechanical physics and the mechanical field, are not relevant in the quantum world. No more than these strong or weak forces so common in textbooks and books for the general public. At most, we can speak about interaction.

2

Two Quantum Studies

In this chapter, I suggest two ways of reading matter. The first reading attempts to reveal the presence of two faces in matter, by analyzing Heisenberg's equation and then suggesting an explanation for the wave function collapse that takes place during measurement. The second reading starts with Schrödinger's equation as its departure point. It is possible to conceive matter from a neohylemorphic perspective with a form–energy duality, what suggests a universal equivalence between information and energy.

2.1. Does the quantum universe observe us?

Quantum mechanics states something that cannot be found in other physics, except by forcing an interpretation on thermodynamics. It specifically explains how information quanta are exchanged. Someday, we will understand that while they were sought to develop a mechanism for particulate and quantified processes, the physicists of the 1920s came upon something strange and unexpected, in a similar way to the discovery of America in 1492. The only difference is that, at the time, physicists did not understand which phenomenological continent they had landed on, nor could they explain which ontological base matter was conceived from.

At the origins of quantum physics, there were two formulations: wave mechanics and matrices. Despite the fact that these are mathematically equivalent, their formulae seem to reveal different things, especially concerning time. In matrix mechanics, it is the observable that evolves through time, whereas in wave mechanics, it is the state vector Ψ that evolves. This evolution can be illustrated by the now famous Schrödinger equation:

$i\hbar \, \partial \Psi / \partial t = - \hbar^2/2m \, \Delta\Psi$ (+ V.Ψ if the particle is subjected to a potential)

and is reduced to:

$$E . \Psi = H \Psi.$$

The "dual" equation in matrix mechanics is formulated with H, which designates the Hamiltonian, and A, the observable introduced as an operator:

$$A(t) = e^{iHt} . A(0) . e^{-iHt}$$

These two equations are interpreted as a representation of the information exchanged between the quantum world and our world. The "contact zone" with our experience world is the quantity observed, that is, the proper value that observable A acquires. What seems enigmatic is that these exact mathematical formulas go beyond conventional physical meaning. We confirm that time does not play the same role in these two equations. In the first formulation, the state vector ψ changes with time, while in the second one, it is observable A that changes.

Heisenberg's equation might lead us to interpret time as associated with the imaginary number i. We may well admit that "real time" is given by the formula i.t. This possibility is not new in fact, it was exposed by Stephen Hawking, for whom the "real" t time in the cosmos might be imaginary and imaginary time might be real. This time reveals the cyclical nature of things, a kind of movement between one world and another world; more precisely, between the world observed and the observing world, or a split between the world expressed and the receptive world. The material physical world of the senses presents itself as an interface between two universes, the one that observes and the one that is observed. It is thus necessary to conceive a double-faced world, a world that accounts both for expression and reception.

First, we place observable A at the interface of these two “worlds”. Then, we place two periodic dynamics on one side and on the other, which are written as i.t and –i.t, which means that what is expressed with a positive temporal dynamic is perceived or received with a negative temporal dynamic. The observable then joins together two complementary processes, the one that expresses and the one that receives (prints). This is a kind of interlock, as occurs with a key and a lock, except that this assembly relies upon the intimate nature of time. For instance, let us imagine a rotating container that holds a certain amount of liquid, in front of another container, rotating in the opposite direction, which is also able to receive the same amount of liquid. When the process occurs, the liquid moves from one container to the other, and we say that a measurement has been made. Actually, the observed system is observant at the same time, and it acquires a quantum of information in the same way as the physicist, who can detect information when he amplifies the signal.

The double-faced phenomenological structure is revealed thanks to the evolutionary equation of the observable: $A\ (t) = e^{iHt}.\ A(0).\ e\text{-}^{iHt}$. This structure can also be found in quantum mechanics as formulated by Hilbert and Dirac. The bra–ket notation leads back to this interfacial structure between an observant and an observed world. Or more precisely, between two faces of the real, two sorts of “substantial fields” whose connection is the interactive place where expression takes place, as well as its reception. This connection is dynamic, of course. The operator embodies the representation of a mechanism that makes it possible for expression and reception to be coupled. There is another notation system that shows this coupling [FEY 79, p. 424]:

$$<\varphi \mid A^{+} \mid \psi> = <\psi \mid A \mid \varphi>^{*}$$

Operator A has a double that we consider as a mirror operator and which is called the adjoint operator or A+ ("A cross"). The term on the left, designated as "bra", and the one on the right, designated as "ket", take part in the complete description of the system, with both faces. The quantum universe observes us at the same time that we observe it. The mechanism involved is a kind of mirror reversal. Not only does "reverse-mirror" symmetry work with operators (see Chapter 1), but it also works with complete representation, including the conjugate expression to the right which, when represented in the Argand plane, appears as the reflected image of the expression to the left of the equivalence. The description of the expressive face depends on positive imaginary time and that of the receptive or perceptive face is associated with negative imaginary time. This is what enables a double dynamic with two opposite "quantum times".

One statement summarizes our point: *in quantum theory, every action is an interaction, whose complete description uses dual, double-faced representation, that of expression or of reception.* This assertion can be inserted in a broader theory of information, because it includes transmitter and receiver. Contrary to Shannon's classic theory where transmitter and receiver were "classical" devices separated in space and time; in quantum theory, these are both intertwined.

2.2. A detour by statistical mechanics

The enigma of imaginary time is also present in the context of statistical mechanics, with a truly enlightening presentation [PRI 96]. In statistical mechanics, the system's description uses the ρ density matrix introduced by von Neumann so as to describe the whole of quantum states in a physical system. This actually makes it possible to combine quantum physics and statistics. The density matrix is an operator that contains information available to be used in a statistical and quantum system. Regarding the temporal evolution of the system, this is given by the Liouville equation, which adopts the following form:

$ih\,\partial\rho/\partial t = [L\rho]$ in which $[L\,\rho]$ represents the H Hamiltonian switch with ρ. The formal solution to the equation can be written in two ways [PRI 96, p. 169]:

$$\rho(t) = e^{-iLt}\rho(0) \text{ or rather } \rho(t) = e^{iHt}.\,\rho(0).\,e^{-iHt}$$

Prigogine considered that the second formula confirmed an observation made by Eddington. It is as if we had two independent dynamic evolutions, one pointing to the future and the other to the past (see also Chapter 10). We will certainly note the Hermitian character of the density matrix, as well as in quantum mechanics operators. This property, together with temporal dynamics, then jumps into our interpretative framework of the physics of expression and reception. Only that this time, the description is external. It is the "quantum monads" of the statistical system that are expressed and "perceived". If we go a step further in Prigogine's demonstration, who was interested in the irreversibility of time, we will discover that while the solution to the equation is symmetrical regarding time, under particular resonances, symmetry was broken. As these (Poincaré) resonances actually couple the two temporal dynamics, this provides us with a clue to the genesis of the "classical" world as derived from the quantum world. Our world is the result of a coupling between expressive and receptive dynamics.

2.3. Expressive and receptive waves in quantum observation

Thus, quantum mechanics describes two dynamics whose understanding was outlined by Prigogine in relation to the enigmas of time. In quantum theory, the conjugate functions Ψ* and Ψ are interpreted as an indirect description of two waves, one pointing towards reception and the other towards expression. The observables' likelihood of occurring is calculated by multiplying Ψ* by Ψ. Following Eddington, which Prigogine interpreted as the meeting between two times [PRI 96, p. 160], in my interpretation, there are two dynamics that come together, expression and reception. This would be similar to a slot machine with two spinning wheels, one with figures and the other with anti-figures, in other words, expression and reception. When the anti-figure comes in front of the figure, an observable is produced. This process is probabilistic, as in the slot machine. Within quantum formalism, probabilities are given by the multiplication of coefficients placed in front of each proper vector. In addition, this is the interpretation provided by Born.

Everything happens as if two wheels were turning, that of expression and that of reception. When figures are complementary, they click and an observable is expressed in relation to the "container" or "recipient". Given a system described by the state vector: a Ψ_a + b Ψ_b + c Ψ_c, with normed coefficients a, b and c (that is, the sum of the squares must be equal to 1). The probability that the observable yields value A corresponding to the vector Ψa is obtained by multiplying a Ψ_a by its conjugate a Ψ_a *. This gives $a^2\Psi_a\Psi_a$ *; and therefore a^2, since all Ψx vectors are normed. Prigogine interpreted this calculation as if the probability resulted from the meeting of the wave going to the future and the wave going to the past. In the interpretation that I offer, it is the expressive wave that meets the receptive wave. This makes it possible to explain the updating process of an observable, that is, the collapse of the state vector, which results from a coincidence where the two waves (expressive and receptive) meet and somehow resonate with each other.

The meeting process of "waves" of quantum substance can be visualized thanks to the image of two slot machine roulettes, the image of a key and a lock, or even a container that is filled as soon as it meets the exact contents it can hold. This quantum process shows similarities with intercellular communication systems, such as membrane receptors, proteins with a sort of "hollow" (the active site) that receives the mediating molecule. It is possible for these molecular mechanisms to be connected (at least analogically), with quantum phenomena. The doublet expression-reception is therefore a universal principle in the field of "material" interactions. *Let us recall the main result of quantum mechanics that there is no expression without reception.* In other words, if we wholly consider quantum dynamics, information circulates in two directions.

Prigogine's interpretation should help us enlighten one of physics' 20th Century enigmas, known as wave function collapse. It is unnecessary to evoke the many pages that have been written on this subject, as well as the attempts to find a new formalism that makes it possible to introduce a causality supplement. It suffices to consider the basic theory and to attribute physical meaning to it.

A microphysical experiment requires a device, as well as a representation of the state of the system before measurement. This representation is provided in the form of a state vector made up of a superposition of proper vectors, such as: $a\ \Psi_a + b\ \Psi_b + c\ \Psi_c$. I interpret this formula as the representation of a superposition of "quantum substance" waves, or expressive waves, to be more precise. The complete representation is obtained by adding the receptive waves associated with the conjugate vector, which is the mirror image of the expressive vector in the Argand plane. This gives: $a\ \Psi^*_a + b\ \Psi^*_b + c\ \Psi^*_c$. In this representation, vector Ψ_a is shown in the image of a whole substance prominence, whereas its conjugate Ψ^*_a is a kind of hollow. The expressive vector swells and the receptive vector "plummets". The state vector collapses when a "swell" coincides with a "plunge".

2.4. Wave function fission

If the results obtained by Humphrey Maris's team are confirmed, then we can say that the fission of the electronic wave function constitutes a major advance in quantum physics. This result was obtained by making an electron interact with super fluid liquid helium. Therefore, a cavity was produced, a kind of bubble in fact due to the electron's effect on helium, but before detecting the electron, other negatively ionized "micro-objects" reached the detector. In this new experiment, Maris and his team attempted to produce an exotic ion by making a photon interact in such a way that it changed the quantum state of the bubble. They found 18 of these negatively charged "exotic objects". Three hypotheses were made, but one is almost certain: the fission of the electronic wave function [WEI 14].

Let us recall that a system is described by a wave function. In order to keep the argument simple, we will analyze a rather uncomplicated function:

$$\Psi = a.\ \psi\ (A) + b.\ \psi\ (B).$$

This function describes a system that, when observed, only gives A or B as a result. In this formula, a and b are coefficients whose squares indicate the probability of producing A or B. Actually, it is quite similar to a lottery. To put it simply, ψ (A) represents a ticket and when we observe A, we win lot A. Quantum observation is like playing a lottery where only one lot is won.

1) The orthodox interpretation was formulated at the Solvay Congress of 1927, where Bohr's conception prevailed. His position stated that as long as the lottery had not yet taken place, we could not know anything about the system. Later, once the lottery (the experiment) was launched, all we could know is which lot won, that is, A or B. Actually, we cannot access the development of the lottery, namely, the game tickets, ψ (A) and ψ (B). Then, once the observation process has taken place, the system is in the quantum state corresponding to the winning lot, with the ticket assigned to it, except that this ticket has no physical reality.

2) Decoherence describes an experiment in which the lottery is launched with such finesse that we can actually observe matter and the way in which the system acquires the two values, A and B. In fact, it is quite strange: we actually win both lots at once because the system works as if we had both tickets in hand. But do not delude yourself, we cannot cheat nature. Actually, to win the lot materially speaking, we should carry out a measurement experiment, play the game until the end and materialize the lot, making the quantum system interact with the observing (classical) system. Now, as soon as this happens, consistency is removed and we only obtain one lot. After all, the experience of decoherence is consistent with the orthodox interpretation, which it does not challenge.

3) The fission of the wave function observed by Maris's team opened an unprecedented perspective provided that the results showed the possibility of separating the lottery tickets in a quantum system,

consisting of an electron submerged into liquid helium. Every bubble absorbed a piece of Ψ or game ticket. From the physical point of view, the bubble created by the electron moved into helium at a certain speed and reached the detector. The type of object was also observed (18 objects in particular), but there might have been infinite objects arriving before the electronic bubble, as could have been the trace of the quantum lottery tickets, ψ (x). These results puzzled the scientific community. Maris was not willing to revisit the basics of quantum mechanics, but preferred to stick to a theory matching with experience. But for others like Michael Byrne, these results threatened to overturn quantum physics, which now seemed to be "walking on its head".

4) In conclusion, the wave function cannot be reduced to a "mathematical object" but possesses "physical qualities", if my interpretation of Maris' experience is correct. In which case, quantum physics is not invalidated, but it is the orthodox interpretation that is surpassed, with an option for the realism of the wave function, but not in Bohm's sense. Because in this experiment, the electron arrived at the detector, and was therefore not "piloted" by the wave function. We will still have to wait for other experimental confirmations in order to be certain about the phenomenon. If it is confirmed, this discovery is more important than decoherence and can be placed at the level of Alain Aspect's experiments on entanglement. Wave function fission depends on helium, because with its few Kelvin degrees, it makes it possible to considerably slow the dynamics of quantum information.

2.5. Form–energy duality in Schrödinger's equation

2.5.1. *Separation of spatial and temporal functions*

The wave function associated with a particle is a complex function defined for any point in space and time. Schrödinger's equation determines its spatial and temporal evolution. This equation should make it possible to calculate the wave function associated with a particle endowed with non-relativistic energy in the presence of a given potential at any point in space. We know that the direct integration of the equation is possible only in certain cases, and particularly when we are looking for stationary solutions. The trick is to separate the spatial and temporal components in the wave function:

$$\Psi\ (r, t) = \Psi 1\ (r)\ \Psi 2\ (t).$$

Applying this function to the equation

$$ih\ \partial\ \Psi\ /\ \partial\ t = -\ h^2/2m\ \Delta\Psi + V.\Psi,$$

we obtain the following formula:

$$h^2/2m\ \Psi 2\ \Delta\Psi 1 + V.\Psi 1\ .\Psi 2 = ih\ \Psi 1\ \partial\ \Psi 2/\ \partial\ t.$$

Dividing all the members of the equation by $\Psi 1\ .\Psi 2$, we obtain this equivalence:

$$-\ h^2/2m\ \Delta\Psi 1\ /\ \Psi 1 + V\ (r) = ih\ (\partial\Psi 2\ /\ \partial\ t)\ /\ \Psi 2.$$

Equality between a spatial function and a time function is only possible if each member is equal to a constant, which in this case we will determine as equal to E (energy in fact). We thus obtain two separate equations, one that must be verified by the time function, and the other, by the spatial function. For the temporal function, integration is trivial by introducing constant C:

$$\Psi 2 = Ce^{iEt/h}$$

The complete solution to the equation is then composed of the result of a spatial function (not yet defined since we have to integrate the equation in order to obtain it) and a time function:

$$\Psi\ (r, t) = C.\ \Psi 1(r).e^{iEt/h}$$

Applying operator $ih\ \partial\ /\ \partial\ t$ to the complete wave function yields a proper value for E, the constant introduced previously. The (apparently arbitrarily) introduced constant is energy. Then, the complete wave function is composed of a spatial function multiplied by a temporal function. Later, it suffices to integrate the equation, by specifying the spatial wave function.

2.5.2. *Solution to the equation and quantum formalization of the hydrogen atom*

The hydrogen atom is composed of an electron and a proton. A mathematical trick makes it possible to obtain two equations, one of which will not only describe the relative “motion” of the electron as regards the proton, but, more importantly, it will determine the shape of the electron cloud around it. Coulomb’s attraction potential is a spherical symmetry expression, since it solely depends on the “distance” that separates the electron from the proton:

$V = -K/r$, where K is a known constant.

Schrödinger’s equation integrates by adopting a spherical coordinate system that comprises the radial parameter r (representing the distance from the electron to the nucleus), as well as two angular parameters. Then, the shape of Schrödinger’s equation is modified. We are looking for a solution resulting from a radial function (dependent on r) and from an angular function (depending on the two angles). This produces two equations that integrate separately.

Solutions are numerous and depend on three quantum numbers: n, l, m [RIV 89, Chapters 1 and 2]. Number n defines an energy level. The value of l is situated between 0 and n-1, whereas m is between 1 and -1. For instance, if n = 3, then l could adopt values 0, 1, 2, and m could adopt values -2, -1, 0, 1, 2. The solution is therefore the product of a radial function dependent on n and l, and an angular function dependent on l and m. We thus perceive the “formal wealth” engendered by the equation. Let us now take a look at a radial function R (r) and an angular function Y (θ, χ), for example:

$$R\,2, 1 = (Z/a)^{3/2}\ 1/2\sqrt{6}.\ Zr/a.\ e^{-Zr/2a}$$

$$Y\,3, +1 = \sqrt{31/8}\sqrt{\pi}.\ \sin\theta\ (5(\cos\theta)^2 - 1)\ e^{i\chi}$$

For each stationary state, there exists a spatial function of the electron whose square is interpreted as a likelihood of presence of this electron, which is not a material point. Each stationary state occupies an energy level which exclusively depends on number n. Thus, energy is associated with several possible forms of the electronic cloud. These energies are calculated with the following formula:

$$En = Eo/n^2;\ \text{avec } Eo = -13.6\ \text{eV}$$

In fact, N number takes part in the radial function. An atom can pass from one state to another either by absorbing or by emitting an amount of radiative energy. If the electron goes from level n = a to level n = b, then the atom emits or absorbs an electromagnetic quantum whose frequency depends on the difference in energy:

$$\upsilon = (Eo/a^2 - Eo/b^2)/h$$

The measurement of the hydrogen atom emission rays confirms the values calculated, thanks to quantum theory. Taking other experimental confirmations into consideration, we can see that this abstract formalism describes the properties of atoms accurately enough (see Chapter 5 on quantum chemistry).

2.5.3. *Form–energy diagram and correspondences*

During an energy transition, the atom changes shape through an enigmatic process. As noted by the founding fathers, quantum theory does not explain transitions between stationary states. During a transition, the initial form is "dissolved", while a new form appears and a quantum of energy is exchanged. The form–energy conjugation can be illustrated by setting up a table that matches energies with radial functions. For a value of n = 1, 2, 3 (with l = 0, 1, 2), the table assumes the following form:

	E1	E2	E3
R1, 0	*		
R2, 0		*	
R2, 1		*	
R3, 0			*
R3, 1			*
R3, 2			*

* This representation illustrates form–energy duality.

Table 2.1. *Form–energy duality*

Each state of the atom uses a certain amount of energy to achieve the expression of a form. What is more, several forms use the same amount of energy (hence, the definition of degenerate states or degenerate energy spectrum). The complete wave function is written as the product of the space function and the time function dependent on $e^{iEt/h}$ energy. This periodic function confirms the vibratory nature of electronic processes around the nucleus. As for the spatial function, it makes it possible to deduce the structure of other atoms as well as the resulting inter-atomic bonds, while it engenders the formal wealth of the molecular world that chemists study (see Chapter 5).

2.5.4. *Form and content in the quantum atom*

The analysis of the quantum hydrogen atom confirms the dual conception of matter that we can explain, thanks to several complementary notions. Each orbital represents an expressed form, belonging to the world of extended bodies and phenomena. Since the atom has only one electron, only one of the orbitals is occupied. This leads us to imagine non-expressed orbitals as shapes, or better still, as encoded, compressed, unexpressed information. In other words, this compressed information set is conducive to the ontophysical (not to say metaphysical) notion of content. We thus have two poles, content and form, or expressed in other words, compressed information and expressed forms. The pole of being and the pole of the phenomenon. Being and phenomenon, these two poles vaguely point to a double theory of matter, already glimpsed by Plotinus and confirmed by the most recent data of contemporary science.

3

Quantum Entanglement

3.1. Some considerations on quantum entanglement

In the quantum world, decoherence and entanglement each delineate a conjecture about the nature of the material universe and the way in which this presents itself at the finest level accessible to the investigation of researchers. Decoherence concerns conjecture about measurement. In the context of the physics of communication, decoherence seeks to elucidate how we move from quantum information (superposition of states) towards the extracted and expressed information of our classical material world. Entanglement concerns a totally different conjecture, that is, the distribution of quantum information throughout several systems described with state vectors in a properly denominated "entangled" configuration. This configuration is one of the most enigmatic challenges for contemporary science.

The famous quantum physics course published by Feynman contains 500 pages sprinkled with mathematical formulas, but there is practically no mention of entanglement. This outstanding detail can be explained through the chronology of events. Feynman published his treatise in 1965, but the fact is that quantum entanglement had only been theorized one year earlier, in 1964, with an article published by Bell, which highlighted the inequalities that made it possible to experiment with entanglement. Bell's article had gone virtually unnoticed because not only was quantum entanglement excluded from the topics taught at academic level, but it was also considered as something external to the experimental field, something like an epistemological tool for working on the interpretation of quantum mechanics. At that time, physicists were going through such a deep state of experimental drunkenness that they neglected the majority of questions dealing with the meaning of quantum theory. It was only after Alain Aspect's experiments in 1982 that physicists really became interested in entanglement as a phenomenon, and that it now stands out in scientific concern, together with the questions about interpretation and, above all, the possible applications in the field of quantum computers. Nowadays, entanglement officially makes part of quantum physics. This can be seen in Leonard Susskind's book devoted to quantum mechanics and which targets a large audience.

What makes Susskind's course singular is the attention dedicated to entanglement. The book's preface thoroughly makes it clear that despite the fact that it constitutes the essential property of quantum physics, this notion was ignored by generations of quantum textbooks. This property might not have been discovered, but was certainly suspected by Einstein and two of his colleagues when he published his famous 1935 article devoted to the EPR paradox, in order to push Bohr back into his epistemological entrenchments. However, the addressee dismissed the article and the case was closed until 1964, when Bell published his inequalities, using spin variables instead of impulse and position, as had been done in the EPR thought experiment. Thus, entanglement added even more strangeness to a quantum mechanics that was already enigmatic.

The unveiling of the quantum world presupposes an art of interpretation because mathematical symbols are yet unable to "speak". In fact, they are similar to the hieroglyphs used by scientists with the intention of describing matter. If an improbable quantum mechanics Champollion existed, he might enlighten us by translating the message that matter communicates to us through the laws and the specialized language used by physicists for describing the quantum world. Physics has always presented itself from a dual perspective, not only as a kind of heuristic user manual for manipulating nature, but more importantly by deploying a special type of language that makes it possible to talk about nature.

3.2. The mystery of quantum entanglement

Quantum entanglement is a property of nature which is not intelligible with the classical or so-called primitive categories of physics, which are widely used for describing the experience of the world (time, space, object, force, etc.). In the English language, the term *entanglement* properly designates a particularly entwined or confused situation. We cannot express it in clearer terms. Quantum entanglement has certainly confused physicists, because this property only appears in a configuration that comprises at least two quantum systems, separately observed. This is precisely the case of the EPR thought experiment with two particles, A and B. Each particle is sent to a detector. While one measures A's position, the other one measures B's impulse, each with maximum precision. By virtue of impulse conservation, the experiment on B makes it possible to deduce A's impulse without observation, which transgresses Heisenberg's uncertainty relation. Entanglement did not stay at the stage of a thought experiment, in fact. It was pursued until properties such as spin or photon polarization were observed, within a rigorous mathematical framework developed by Bell. Now, entanglement indicates how quantum information is organized in combined systems.

Let us examine a classically combined system. For this, I would like to work with the example offered by Susskind. Bob and Alice are two characters and each of them represents a system. Charlie has an urn with a red ball and a green ball. He randomly takes one ball, puts it in a box and gives it to Alice. He then puts the other ball in another box and gives it to Bob. Later, Alice goes on a long journey to Uranus. Bob and Alice have synchronized their clocks. When Alice arrives on Uranus, she opens the box and finds that it contains a green ball. At that same moment, she knows that Bob will find a red ball in his box. However, this does not mean that Bob instantly communicated the information to Alice. In fact, this would be impossible because the transmission of information in the electromagnetic range is not instantaneous. Now, let us assume that Charlie painted "+1" on the red ball and "−1" on the green one. And that he repeated the experience a thousand times. Alice and Bob wrote the results down. Charlie calculated the σ average and found zero for A and B. On the contrary, when he multiplied the two figures obtained in each experiment, the product was −1. This means that <σA> x <σB> is equal to zero and is then different from <σA x σB>, which equals −1. This difference is a correlation coefficient which, in case of non-nullity, indicates that the two systems are correlated, or entangled, but always in the context of a classical situation, what is not strange at all, and is subject to basic probability calculation. This correlation is related to the fact that Charlie has two balls. Then, complete knowledge of a correlated "classical system", comprising two parts, implies the complete knowledge of each of the parts. This is similar to the thorough knowledge of a motor vehicle whose every single part is perfectly known by the mechanic. However, when we take an entangled quantum system into consideration, this principle that connects the totality with the parts is no longer valid. As we will see, in the quantum world, we can completely know a system without knowing anything about each of its parts.

Now let us imagine Bob and Alice as two quantum observation systems. Instead of taking balls or numbers, we use a quantum observable, the spin. Now, let A be Alice's spin and B be Bob's spin. Each of the spins can be observed in an up or a down state. The combination of the two systems is described by the tensor product of A's spin vector by B's spin vector. We thus obtain a base of four proper vectors defined in this way: uu>, ud>, of>, dd>. For example, ud> means that Alice's spin is up, while Bob's spin is down. A spin prepared by Alice is described as a combination of both vectors, u> and d>:

$$\{A\} = \alpha_u \, |u> + \alpha_d \, |d>$$

The same goes for Bob, whose preparation is thus defined:

$$\{B\} = \beta_u \, |u> + \beta_d \, |d>$$

The composite state jointly prepared by Alice and Bob is designated as a product state (or simply separable state) obtained by multiplying the tensor product of both vectors:

$$| \textit{product-state} > = (\alpha_u |u> + \alpha_d |d>) \times (\beta_u |u> + \beta_d |d>)$$

The product state describes two systems that are indifferent to one another other. If Alice carries out an experiment on her prepared system, whatever the outcome, it will not affect Bob's experience. Everything happens as if Alice and Bob ignored each other. However, the situation may change if we admit that in the product space we can build state vectors at a much higher number than in the product states defined previously. In other words, the state space is far richer than the set of product states formed by the independent preparations of Alice and Bob. Something new is going on and this is what we call entanglement [SUS 16, p. 161]; with more or less entangled states. The singlet state is an example of a state prepared with maximum entanglement. Its vector in the product space is written as follows:

$$|\textit{sing}> = 1/\sqrt{2}\ (|ud> - |du>)$$

Now, before we can explain the strange character of entanglement, let us make a preliminary remark. Classic entanglement can be explained, thanks to objects and systems associated with common notions. The experiences with Charlie, Bob and Alice are achievable because these are incarnate individuals. In the quantum world, entanglement does not refer so much to objects, but to the preparation of an experiment with states involving abstract spaces and complex vectors. Navigating from the classical world to the quantum world requires intellectual gymnastics. In fact, Einstein did not discover quantum entanglement in 1935. The EPR experiment was no more than a thought experiment, combining quantum observation and classical conservation laws. It is precisely for this reason that Bohr rejected it. Nevertheless, the EPR experiment had a heuristic value for better understanding quantum entanglement (conceived with two quantum systems), without making reference to classical laws, except for the principle of conservation, which governs all physical phenomena.

Now, let us go back to classic entanglement. It occurs because Charlie uses two balls, and when he gives one to Alice, he gives the other one to Bob. We could as easily imagine a situation without entanglement. Charlie has two balls, gives one to Alice, then comes back with two more balls and gives one to Bob. We understand that there is no correlation in this second option. In the case of an entangled combination as $|\ \text{sing}\rangle$, correlation takes place. With the calculation of quantum probabilities, it is possible to obtain the correlation coefficient given by the following formula:

Correlation: $\langle \sigma A \times \sigma B \rangle - \langle \sigma A \rangle \times \langle \sigma B \rangle$

For a product state that corresponds to an independent preparation by Bob and Alice, this coefficient is zero. For a maximum entanglement state such as the singlet, the coefficient is equal to −1. For partial entanglement states, it is non-zero, for example −0.3 or −0.96. It is this type of calculation that Bell used, and which led to those famous inequalities. Then, it is sufficient to carry out the experiment. After 1982 and Alain Aspect's publication, entanglement became a property of nature and, more specifically, a property that concerns the distribution and organization of quantum information.

Interpretation of entanglement. The first results regarding entanglement led to a state of "learning effervescence", not only in the scientific community but also among epistemologists and other relatively disoriented philosophers. Erroneous assertions were made, often in articles for a non-specialized audience. New notions were used (non-separability and non-locality), but without further clarifying the subject. Contrary to what was said in popular presentations, Aspect's experiments were not successful in arbitrating the conflict between Einstein's classical position and the validity of quantum mechanics. In fact, they made it possible to verify a new property that emerges when we study *not one, but two* quantum systems. Therefore, are entangled states a formal invention related to quantum mathematics or are they a property of nature unveiled by quantum experiments? Experience provided an answer: entanglements are a "property". To be complete, entanglement does not require invoking signal transmission at a superluminal speed. Entanglement reveals a property of nature, but nothing else. To imagine superluminal signals is to remain still attached to a classical image of things, in which everything is based on interactions conceived in terms of scope and communication. Entanglement actually describes how quantum information is "arranged" in two quantum systems separated by an extension.

Susskind interpreted entanglement by means of a witty joke that reverses the common sense of our everyday life. In an entangled system, we can know everything about the system without knowing anything about each part. In the ordinary world, this would be like imagining a mechanic who knows everything about a vehicle but who knows nothing about its wheels, the cylinders, the chassis, the carburetor and the transmission. This picture shows to what extent the quantum world requires an extraordinary way of thinking. This was already suspected in 1927 and the situation only became more acute, because entanglement revealed something new and unexpected for the modern mind. Science barely glimpsed the astonishing consequences of this property of matter, which, when unveiled by quantum physics, could explain many other things. For, let us not forget, the cosmos is full of matter and we, living creatures among others, are also composed of matter arranged in a certain way.

The place of *entanglement* can be partially clarified if it is framed against the separation between the physics of communications and the physics of arrangements. So far, we have understood that the two quantum observations *are not involved in (remote) communication, but that they reveal something about the arrangement of information, not of masses*! Then, we can also understand why, on the basis of quantum information, entanglement is used in the context of cosmology for explaining the way in which the parts of the extension are, so to speak, arranged thanks to a quantum gum (see a presentation in [DUG 17], Chapter 5, section 2).

3.3. Quantum entanglement and complex sciences: information and system

Quantum physics applied to a system describes two things. First, we have the state vector, which describes the necessary information contained in this system and which makes it accessible to experience when transferred to the laboratory world. Second, information can be extracted in the form of observables, whose calculation is made when operators are applied to state vectors. Entanglement concerns two devices with additional complexity, contained in three types of representation: factorizable product states, entangled states and partially entangled states.

Now, is entanglement an isolated physical relationship case occurring when a scientist prepares two systems in an entangled state, or is it a fundamental property of nature? In which case, entanglement would be largely present. A strong hypothesis would point that entangled systems could be the rule, while independent systems would be the exception. Some specialists do not hesitate to examine this hypothesis which, if legitimately proved, would make it possible to configure the greatest scientific questions with new elements, among which one stands out because of its being simultaneously unresolved and universal. This is the central question of systemics. How can we explain the working of a whole whose properties cannot be conceived as the sum of the properties of each part? And given the fact that systems evolve in time, this question also concerns the emergence of a new system, as if it were a new species or a work of art. We can observe that while entanglement has remained outside most scientific fields for decades, now the situation seems to be changing.

Systemic research posed the central question about emergence through the generic concept of self-organization, which was largely present and reached a climax between 1970 and 1990. Simultaneously, this era of intellectual effervescence was marked by the theories of chaos. According to systems theorists, the outlines of entanglement could be explained in at least three different ways. The phenomenon was first considered as a curiosity, with some hesitation about its existence and its extreme strangeness, which prevented it from becoming a central concept for understanding nature. Second, unlike the classical, strong, weak and especially electromagnetic "forces", together with atomism and all the chemistry that derive from this approach, entanglement might not be the result of interaction. On the contrary, entanglement could be conceived as a formal relationship between two parts of nature with no interaction, while all modern science has been based on interactions (including systemic conceptions such as autopoiesis, with its concept of operational closure as proposed by Varela). Third, genetic advances carried out during the 1990s guided systemic thinking in a cybernetic-inspired direction. In genes, biologists thought they had found the ultimate solution for understanding living creatures, and were intoxicated by the prospects opened by the sequencing of the human genome.

Modern science is based on classical notions: interaction, object, reaction, force and especially mechanism. What is specific about entanglement is that it features a situation that goes beyond the modern frames of scientific thought, and it is precisely this feature that keeps "Cartesian" professionals apart, whereas the curious and adventurous minds are ready to hear what resonates through quantum entanglement. But before understanding can be reached, it is important to properly plot the conjecture.

Entangled systems can be prepared by using several devices and playing with elementary properties such as electronic spin or photon polarization. Ion pairs can also be configured in an entangled state, using a highly elaborated device. Let us take the example of beryllium ions and a pair of magnesium ions separated by a quarter of a millimeter. These pairs oscillate because of electrostatic repulsion. By using electric fields and laser impulses, these two pairs can be placed under a composite entangled state. No surprise so far, unless we question reality. Are entangled systems phenomena produced by science under special conditions or are they rather fundamental properties of nature? The second option might be possible, but it would lead us to envisioning ordinary quantum experiences as something produced by a science that tends to model disentangled situations, by separating components. After all, disentanglement might lead to nature's free choice and independence, whereas entanglement is a property that takes part in the order of things, while it shapes classical worlds. For the time being, we are naked against the quantum "epistemological wall".

I have perused the colossal arXiv library for articles containing the word entanglement in their title, associated with another keyword. By combining with *entropy*, we obtain more than 800 responses. With *gravity*, there are about 450 responses. With *complexity*, it falls to 45 and, finally, no article combines *entanglement* with *biology* in its title, whereas combining it with *DNA,* we obtain six very recent results. This means little (or poor) advances in the field of entanglement applied to biological systems. Nevertheless, for the time being, entanglement remains a notion belonging to the world of physics and that closely concerns entropy, as well as the formulation of quantum gravity. The association with entropy brings no surprise at all, if we admit that entanglement determines a certain configuration of information related to quantum states. What really seems surprising is the small number of articles devoted to complexity, not to mention the complete absence of studies on entanglement applied to biological systems, except for some five articles, this time identified with a query made on the text and not the title. So far, quantum entanglement does not seem to be currently considered as a phenomenon that could help us explain the complexity of living systems.

After 50 years of research in complexity theory and 30 years of physical studies on entanglement, we find very little investigation about a possible role of entanglement in the order and the information of complex systems. Modern biology thinks that it can find the explanation for the origins of life, evolution and the emergence of consciousness in the complexity of mechanisms, but this science is now going through an impasse. The observable complexity is not the cause, but the consequence of more fundamental processes. Only radical speculative advances such as those promised by quantum physics can spur forward the understanding of complex things. May these few lines inspire generations of young researchers to explore this path. The EPR experiment conducted by Einstein was a thought experiment. Science also moves forward, thanks to thought experiments! Entanglement has everything needed to help us think about the missing links in the explanation of life and consciousness.

3.4. Concluding remark about strange information

Let us insist once more on the fact that quantum physics describes processes that are incomprehensible from the classical point of view, which professes the separation between observer and object. These processes are conceived as extractions of elementary information that go across domains: that of quantum states and that of observables. The description of this extraction is partly described by decoherence procedures. This science tells us that nature and matter contain much more information than we can extract, at least in a limited lapse of time.

Moreover, quantum information is structured and distributed in a very particular way when configured in entangled systems. Information acquired in one place provides information in another place, without need of measuring it. Thus, the formal universe of quantum information is both redundant and structured. When this universe communicates, it is a two-way process, so that quantum information is modified as soon as it comes into contact with the world of information expressed in the objectified world. Quantum mechanics does not fully grasp the processes related to quantum information, though. We must go through the phenomenal interface in order to obtain the signals that help us understand surface matter.

Basic quantum states contain elementary information that is carried from its material dimension to the laboratory, for observation purposes. If we now notice that entanglement makes it possible to prepare states in very large numbers, this means that the world of quantum states is not fixed, but presents itself as a reservoir of quantum information that can be combined in order to create other pieces of information. Quantum physics has partial access to the creative dynamism inherent in fundamental matter.

Finally, a hypothesis on entanglement contains the possibility of returning to a universal cosmological time which would not be that of Newton, associated with the extended scene, but produced by arranged quantum matter. The big explanation between quantum theory and relativistic cosmology is under way, but not in the way suggested by Einstein.

4

Quantum Darwinism and the Macroscopic World

Quantum Darwinism is a reflection about the articulation between the world described by quantum physics and the macroscopic world. For some, it is a description of nature, while for others, it is an epistemological curiosity. However, this reflection seems to possess heuristic interest for understanding the emergence of a macroscopic world made from a primordial matter of quantum information.

4.1. Quantum Darwinism, from quantum to the observable world

Newton's physics separates ontology from nomology and this, following the principle of the states of the world, exists regardless of the information that scientists can obtain about these states. Quantum theory abolished this separation and brought us back to realism. Wojciech Zurek, one of the founding fathers of "quantum Darwinism", expressed it in one of the latest articles devoted to the subject [ZUR 14]. One of the "implicit" challenges of quantum Darwinism is to explain how, starting from a quantum description of the irreducible nature of our classical universe, this same classical universe emerged from the quantum world that we can observe. The physical link between quantum and classical is particular, indeed. This can be sharply outlined thanks to decoherence, a phenomenon whose experimental study was first successfully carried out in the 1990s. Decoherence theory, of which Zurek is also a founding father, had been developed in the 1980s and conceived in 1970 by Heinz-Dieter Zeh.

While decoherence does not explain the collapse of overlapping states, it is sometimes used as an answer to the question concerning the phenomenon of observables, by explaining why we observe "states". Zurek repeatedly insists on information in the quantum universe. And it is this themata, in the sense of Holton, which will guide our reflection on quantum Darwinism, questioning how the "world of quantum states" communicates with our macroscopic world and the phenomenal forms that we perceive, while we observe the quantum forms in devices.

Louis de Broglie, one of the founding fathers of quantum mechanics, had already attempted to establish a connection between the two physics. He conceived the idea of a pilot wave that would reintroduce causality in quantum physics, but despite his efforts, he was not able to connect both physics. While classical physics uses primitive notions (such as space, forces and masses), quantum physics uses specific notions belonging to the vocabulary of specialists. This vocabulary has not remained fixed however, and neither has quantum theory, so, as Zurek pointed out, new concepts with theoretical and empirical implications appeared well after the 1930s. Among these, we can mention Bell's theorem about quantum entanglement, decoherence, and the theorem about the impossibility of quantum cloning. This theorem, formulated by Zurek and his colleagues in 1982, shows that we cannot duplicate an unknown state because in order to do this, we would need to be able to measure it, which would make the system collapse to one of its own states, and would result in a loss of information about the state we are trying to clone. This theorem places the notion of information at the center of quantum physics. Moreover, this piece of information accentuates the schism between the classical world and the quantum universe. In fact, in the classical world, cloning is easily achievable. It is what I do when I reproduce quotations from scholarly works. Also, this is exactly what happens in a banal photocopy machine, or a printing press or a vinyl record factory. Information is replicated or reproduced, but it is not cloned in the quantum sense of the term.

Philologists know that the words that make up a language evolve with human experience and the understanding of the world. This was true during the transition from the Homeric era to Plato's classical Greece. Homer's Greek was not the same as the Greek of Plato or Sophocles. This phenomenon of "linguistic drift" also occurred from Vedic India to Vedanta India, "term" meaning "achievement". Two founding texts were involved: the *Upanishads* and the *Bhagavad Gita*. The Greeks invented "physics" with Democritus, Anaxagoras and other pre-Socratic thinkers. Plato and Aristotle left detailed writings on nature, with notions derived from the experiences they had about nature (intelligible and sensible, Idea, *eidos*, *hylé*, form, entelechy, etc.). Hindus invented unprecedented ideas such as *purusha*, *prakriti* or the three qualities of energy known as *gunas*. We may think that the novel concepts of quantum physics are words that constitute a new language for understanding matter. Quantum entanglement, decoherence and "non-cloning" are notions that do not appear in the physics of the founding fathers (Zurek). These notions make it possible to develop a "*new physics of information*". Science generates epistemological or even ontological break offs when new concepts appear. Such was the case with force, acceleration and mass - the field for modern mechanics. Quantum physics is no exception to this rule. Nor is biology, a discipline that was shaken by the notions of natural selection and gene.

Zurek forged neologisms so as to introduce decoherence in the context of quantum Darwinism. The notion of *einselection* designates a "mechanism" by which the classical environment classifies quantum states accessible to observation, these states being defined as *pointer states*. When the quantum experiment delivers a result, the phenomenon is interpreted as a transmission of information that uses a communication channel which connects the world of quantum states described in Hilbert's space with the classical macroscopic world, which, according to the hypothesis of "selective sorting" only leaves "classic islets", or *pointer states* that are properly registered in "our space". These exist in the universe in a redundant form, with multiple copies. For Zurek, the classical environment does more than produce decoherence (state vector collapse). This environment is understood as a communication channel that goes beyond the framework used in Shannon's information theory. Besides, it is difficult to define information at the quantum level without risking confusion with the classical information that we find in statistical thermodynamics or in Shannon's theory. This detail did not escape Chris Fields' attention when he delivered the critical review of an important work by Ruth Kastner on the transactional interpretation of quantum physics. In this book, the notion of information was relatively avoided, while it is a relevant notion. The community of physicists was confronted once again, against the information wall.

4.2. The controversy between W. Zurek and R. Kastner

While the project of quantum Darwinism was rich in meaning and presented encouraging perspectives, it was nonetheless fragile in its presuppositions and this is what Ruth Kastner intended to prove by detecting certain circularity in Zurek's reasoning. This means that the explanation of a classical world built from a "quantum universe", does not work when we apply quantum Darwinism: the reasoning is circular, that is, the conclusions that should result from physical reasoning are implicitly introduced from the departure point. More precisely, classical islets or *pointer states* do not emerge from quantum, but derive from classical propositions assumed from the start [KAS 14b]. Kastner goes so far as to compare the circularity of quantum Darwinism with what might also be present in Boltzmann's theorem, where inferred irreversibility is already introduced into the assumptions about molecular chaos.

In order to fully understand the challenges of quantum Darwinism and its limits, we must begin by unraveling the epistemological knots of quantum physics, and try to separate the two puzzles that have emerged. First, we have the phenomenological enigma of quantum measurement. How does the quantum phenomenon occur in our classical world? In other words, assuming that quantum mechanics describes the most basic part of our universe, how does this part appear in the Whole that comprises our environment, our spatial and perceptive extension? Then, we have the ontological enigma. How is a classical universe built from quantum elements? This second enigma is systemic: how do certain elements make a Totality emerge? Indeed, we may wonder whether the quantum particle is a part, a stable component, or just a phenomenon observed in the experimental field. If we do not bear these phenomenological and ontological questions in mind, we will no doubt run into this wall, and it must be this impasse of quantum Darwinism that Kastner tried to explain to us. Incidentally, Kastner wrote an update in the arXiv so as to make it clear that Zurek did not wish to mention his critics in his last article concerning quantum jumps.

I will now use the image of ink and ceramics. In the quantum world, the description of states is represented by superimposed ink drops endowed with movement. Quantum states are dynamic states. The appropriate image for representing the superposition of basic states is that of a palimpsest which is permanently colored with ink, and then fades. The problem of measurement is to understand how the information written on the ink drops is spread out in order to make a stain on the ceramic pieces that represent our classical world. How do we go from the "book of fundamental states written in quantum ink" to the ink drop that produces a stain on traditional ceramics?

Kastner describes quantum Darwinism by means of a triplet (S + P + E). S denotes the system, that is, the book made up by the drops of quantum ink and which remains indefinite, because it contains all possible quantum states. E refers to the environment made of ceramic. P designates the classical *pointer states*, granting the transmission of information from the (S) quantum book to (E) classical ceramics. These (P) *pointer states* have a double character, quantum and classic; then, they are selected by the environment. To continue with the metaphor, everything happens as if the ceramic chose which quantum ink characters must be read and passed to the classical world.

Decoherence makes it possible to detail what is happening. From our classical world, an experimenter introduces a tiny (*mesoscopic*) ceramic needle into a "quantum system". This needle is capable of doing two things: turning the pages of the quantum book and then putting the book back into its place and making a copy of a single page. Decoherence experiments have succeeded in producing an extreme situation in which the needle turns a quantum page, reads it by capturing the ink, then rejects the ink so as to open another page and read it, the whole, in a short period of time. The system is maintained in both states. This is what Serge Haroche observed by introducing previously "prepared" dual-state rubidium atoms into a micro cavity crossed by EM waves. When the cavity exchanges enough energy with the environment, one of the states vanishes and the other becomes "definitely" observed, fixed in the traditional ceramics, in a similar way to a photograph that has become indelible. Quantum ink somehow imprints our phenomenal matter. The quantum book expresses itself in our extended world. It is in this way that we communicate with quantum matter. The fundamental result of decoherence is to have shown that the expression of a state is joint with the suppression of another state. This principle can be extended to any wave function collapse process, which we can also define as the suppression of potential communications.

Quantum Darwinism is a philosophy of nature. Additional assumptions can be introduced, such as the environment selection of *pointer states* and the presence of redundancy in the "encoding" of *pointer states*. This redundancy resembles that of ants "copied" by millions, thanks to the species selection made by the environment. The goal of quantum Darwinism is to find an explanation for the transition between the worlds of "quantum" and the "classical" world. How do we go from the ink of the quantum book to the classical world made of perceptible structures in ceramic?

For Kastner, the division of the Quantico-Darwinian world into (S + P + E) is considered arbitrary because it incorporates too many classical elements, which engenders certain circularity when explaining how classical things are produced by quantum things. Classical observability is introduced from the beginning and is not inferred from the quantum world; in other words, it appears in the conclusion, because it is already present in the argument's premises. The observable world is based on the existence of ceramics, with a redundancy of classical information. Then, the principle of state selection by the environment is represented by a coherent mosaic that enables the experimenters to capture the pieces of these classical ceramics, which present certain regularities, be it in Paris or in Beijing. The pieces of ceramic reveal the quantum ink emanations sorted by an environment whose observation makes it possible to analyze such pieces, whose resulting shape is like a mosaic.

Kastner's analysis is very technical. It is impossible to prove the redundant encoding of the environment exclusively using quantum principles. It is impossible to infer decoherence from the use of matrix S and an equation describing a unitary dynamic (which maintains the sum of probabilities). Quantum Darwinism seems to be based on a mathematical magic trick with the *pointer state* as a connecting element between quantum ink and classical ceramics. The program of quantum Darwinism relies on the questioning of the extraction of quantum information with two conjectures, a phenomenological one (which has become conventional and related to measurement) and an ontological one (related to the constitution of the physical world). It seems that the ontological conjecture cannot be solved if we start from the quantum world. In other words, quantum theory describes communication, but is incomplete and cannot explain how these communications build a "world", if we assume that the world is assembled from quantum elements. And it is this deficit which has to be filled and that came to light when I separated the two physics. In order to explain the macroscopic world, quantum matter (*communications*) is not sufficient; it is necessary to incorporate specific elements and principles of physics of *arrangements*.

4.3. Understanding our classical perception with quantum physics, an anthropomorphic approach

Now, there is another question: how do we perceive the world? This is one of the key questions for both science and philosophy. Kant has suggested some hypotheses in his transcendental aesthetic (*Critique of Pure Reason*). The debate is currently open and refreshed, thanks to the tools of quantum physics. It is from this viewpoint that, following Roger Penrose but in a distinct context, Thomas Durt suggested using quantum physics to explain the perceptive phenomena of our brain; in other words, it is a question of explaining how the non-classical processes of the quantum world play a role in biological systems. This reflection is an extension of quantum Darwinism, whose intention was to connect "quantum matter" with "classical materiality", except that in the anthropomorphic approach, the challenge was to connect "quantum matter" with the perception of the neuronal subject, composed of "living matter" [DUR 10]. Among others, decoherence seeks to describe the communication of quantum information, later transformed into classical information. In the case of the brain, perception makes it possible to observe classical forms from a classical world. And this implies an additional level of difficulty to involve quantum physics in the perception process. In both conjectures, the question involves phenomenology: the phenomenon measured in the laboratory, the phenomenon perceived by the brain.

Durt's reflection deserves a detour, analyzing three different aspects. The first one is the transposition of Zurek's epistemology to the biological domain, assuming that the environment selects, sorts and adjusts the quantum world to the classical world. Except that it is the brain that operates, not laboratory "matter". The brain sorts quantum information out in a similar way as in the material environment of the laboratory. This hypothesis is consistent because the brain is made up of the same matter as that of the laboratory, except that this matter is alive, and subjected to the rules of physics. The second part concerns quantum entanglement. As Durt smartly pointed out, entanglement is most often the rule, and when two quantum components are entangled, then "Cartesian knowledge" collapses. In other words, we cannot access knowledge by juxtaposing the knowledge of the parts. Each component loses its individual character because it is concerned by another component, with which it is intertwined. Entanglement thus paves the way for a holistic conception of "matter" at its most fundamental level. The third observation concerns the role played by entropy, both in the environment's process of selection of classical islets and in the brain's selection of classical information, according to Durt's hypothesis. Nevertheless, entropy remains a descriptive or even heuristic tool, but it does not explain the substance of "physical things" or of "alive, perceptual things".

The author stated the difficulty of explaining how a phenomenological scene is construed with a brain made of matter, a part of whose properties is described thanks to decoherence and entanglement, as well as a hypothesis concerning information selection (this principle being inspired by Darwinian natural selection). He also added this astonishing suggestion: our perception might have become traditional because we might have gradually become "blind" to entanglement. In the end, we will conclude that in the quantum universe, knowledge is a complement of mystery. Durt surreptitiously slipped a colleague's remark in his article, regarding ancient civilizations, which might have had a deeper knowledge of entanglement in the universe. This is indeed possible if we refer to Heraclites' concept of non-apparent adjustment or Anaxagoras' notion that "everything is in everything", as well as his principle of intelligence, which provides order to the small elements thrown pell-mell into nature.

The transition between the quantum world and the classical world is conjectured on the basis of an inquiry around information. We are still lacking a "new physics" which can account for this articulation. This should be a physics of the layout of information in a classical world, applicable to animals or to the human experience. Maybe this physics will be elaborated as a metaphysics, or even a new informational theory of gravity and magnetism; after all, these two guiding principles make our "material and sensible perception" of "objects" possible, not only by means of extension but also thanks to the arrangement of objects with the help of cosmological gravity. *While gravity rules arrangements, does magnetism rule communications*? This new physics will be explored by examining the proposals of Paul Davies, who reflects on the link between information and biology. The complete image should also include time, because the book of our perception opens and closes at every single moment. The present world is shaped by this book, which not only expresses itself, but also imprints itself, with two determinations of time (positive and negative), the present being determined by the past, but also by the future. *Kronos* and *Telos* refer to these two determinations (see Chapter 10).

4.4. From quantum substantial matter to spatiotemporal extension

The quantum book opens and closes, with pages that are constantly being read and unfolded in the world of extended ceramics, and then closed and folded into the world of the never-ending quantum book. Particles and processes are part of a spatiotemporal scene whose description requires a cosmology. Thermodynamics questions the arrow of time and cosmology describes spatiotemporal dynamics, while restrictive relativity concerns the structure of space and time adapted to communications. In fact, there are two space-times. (1) There is Euclidean time, described with a 4D vector by Minkowski, Lorentz, Einstein and in which Maxwell's field "is floating"; (2) there is a time that is curved, dynamic, described with tensors by Einstein and in which the masses of the cosmos "are floating". Finally, (3) there is time appearing in statistical physics, making it possible to describe irreversibility, that is, the impossibility of passing a film upside down by natural means. These descriptions (1), (2) and (3) belong to the classical world. This is also our common world, the one that enables "monads" to communicate, whatever their degree of perception.

In the classical world, a single space-time is used for describing the state of a system. In the quantum world, two particles require a description in 6D. And n particles require a description in 3N dimensions. Each particle seems to read its own quantum book, whereas the classical world is made of a common mosaic in which we communicate. Where does this coherent reading come from? Maybe particles are not elements interpreted in the sense of constituents which deliver a basic description of reality. Indeed, quantum physics is tearing our certainties apart, and this is good news.

Going from quantum physics to extended physics requires going beyond the limits of orthodox interpretation and seeking to explain, not just to describe the appearance of the phenomenon. It seems difficult to make a space emerge from a description that leads to "dissolve" this same space. Nevertheless, some people think that quantum physics contains a theory of space-time and what would ultimately complicate things, since a third formulation of the extent would then be available, or rather a formulation of the specific extent of the quantum world. It is this idea that comes out vaguely from the transactional interpretation invented by Cramer. In this diagram, quantum processes are compared to a transaction conceived as a meeting (a resonance?) between two waves, one delayed and one moved forward. These two waves are designated with a semantics specific to the philosophy of the transaction: we have the wave offered and the confirmation wave. When the confirmation is in "conjunction" with the offer, then an actualized quantum process occurs, in other terms, a transaction takes place.

According to Kastner, it is possible to infer several characters about the emergent extent, starting from the "quantum world". Space exists only when time exists. From a transactional point of view, the concept of space does not make sense if there is no temporal relationship. Conversely, movement in time can be conceived without motion in space. What is more, time in the quantum sense only applies in the realm of the actualized processes and modifications can occur in the "quantum substance", independently from any temporal reference [KAS 14a]. The quantum world thus takes part in a sort of pre-constitution of space and time. In other words, what this world might share with that of relativistic cosmology and special relativity is the property of shaping and/or adjusting the scene on which phenomena deploy. However, the extent of the quantum world is Euclidean, while that of the relativistic cosmos is not (this aspect is what makes relativity irreconcilable with quantum theory). The theory of advanced and delayed waves refers to a pre-formation of space-time, a pre-figuration elaborated in "quantum substance". In the end, quantum physics alone cannot explain the classical world.

Quantum Darwinism concerns the communication between the world of primordial information described by quantum state vectors and the world of classical forms. How do we extract quantum information, how is it transformed into signals leaving traces in the laboratory? This questioning assumes nature conceived as a cosmos having selected communication structures while simultaneously regulating them (see Chapter 7). In the context of quantum Darwinism, it is quantum elementary information that is transmitted and that becomes accessible to experimental capture. *Pointer states* would then be mesoscopic intermediate elements (see Chapter 11) selected by nature so as to communicate and transmit information from the quantum basic states to our phenomenal world. Quantum Darwinism thus presents a heuristic interest for having access to the universal principle of semantic Darwinism and the selection of signals or communicating at different levels by means of hardware interfaces. Because in our world, nature barely selects signals and we, human beings gifted with language, do exactly the same.

5

Chemistry and Quantum Communications

5.1. Brief epistemological notes

While quantum physics describes matter at the elementary level (defined as microphysics), our world is modeled on another scale of magnitude (defined as macroscopic). Twelve grams of carbon contain a number of atoms denoted by a 6 followed by 23 zeros (this is Avogadro's number). The sciences that study chemical properties are different from quantum physics. Chemistry studies the molecular substances that can be transformed. An encyclopaedist wishing to order sciences according to their increasing degree of complexity would place thermodynamics first, second chemistry, third biochemistry, cell biology, physiology and finally, neuroscience. However, this classification is based on an ascending epistemology, not to say reductionism. Classical physics describes systems by representing them with material points that describe solids whose components are stable and indifferent to each other. Now, in chemistry, the transformations of molecular components are studied and these components are no longer indifferent, they interact, as everything does in condensed matter. Biochemistry describes macromolecular structures composed of a very large number of organic molecules, sugars, lipids, amino acids, etc. Cells are integrated by biochemical elements which form structures and which are able to organize circulating information. Finally, organisms are composed of cells in large numbers.

Many sciences hide the quantum description of matter and content themselves with the representation of living materials, substances and systems as a set of components endowed with qualities, and creating systems with new properties when they combine. The questions of morphogenesis and emergence, in particular, remain unanswered. But with the new interpretation of quantum matter, it will be possible to formulate these qucstions in an unprecedented way. In thc ficld of chemistry, quantum physics has been applied with remarkable success and it is this discipline, far more than thermodynamics, which offers a fine understanding of the phenomena of association, bonding and transformation. Quantum chemistry explains the structures of atomic bonds, whereas chemical thermodynamics is more interested in energy balances and reaction speeds. The central concept of quantum chemistry is the "electronic cloud", also referred to as "electronic orbital". In the case of hydrogen, this can be calculated mathematically by solving Schrödinger's equation. By convention, every atom has its electrons placed on the orbitals calculated for hydrogen, and the model works fairly properly! Quantum chemistry is developed as a physics of the forms, and even a physics of communication. In general, thermochemistry tends to ignore the quantum description and uses classical state variables, H enthalpy, S entropy and G free enthalpy. The main goal of classical chemistry is to establish the possibility of reactions, associated energy exchanges and kinetics. Classical chemistry sees things on the surface, imagining atoms as the pieces of a game of Lego reacting in concert.

5.2. Chemistry, a little-known science that studies bondings and reactions

While physics fascinates the general public with its black holes and its big bang theories and the science of genetics appeals in virtue of the manipulations that it makes possible (without setting the question of evolution, aside), who knows chemistry and becomes interested in its theoretical aspects? In fact, chemistry is practiced by hundreds of thousands of scientists around the world, whereas approximately a hundred million molecules are listed, some of them better known because they are common in living systems and sometimes used for manufacturing materials or for therapeutic purposes. Nobody ignores the existence of glucose, of PVC or aspirin. Chemistry works very well and is extremely effective in the industrial field. It also offers many therapeutic tools for medicine; finally, it provides an access for understanding the living on a mechanistic level, which nonetheless remains superficial. Molecule study is essential for grasping the different macroscopic emergences which, after all, are the chemical reactions of matter and rule the transformations in life.

As a science, chemistry remained little known to the general public and besides, the philosophy of chemistry is a very recent discipline, barely a decade old. In *Stanford's Philosophical Encyclopedia*, we can find a thorough study conducted by Michael Weisberg, Paul Needham and Robin Hendry. The molecular level is situated between two epistemological and ontological fields. On the one hand, the sub-molecular component offers microphysical descriptions of chemical bonds which use quantum mechanics. This discipline is known as "quantum chemistry". The other component, the most frequent one, is that of conventional chemistry, which uses the classical tools of thermodynamics. This scientific configuration leads to a sort of epistemological separation where the molecular is drawn towards an intimate understanding of processes involving the quantum universe, and on the other hand, towards the classical and macroscopic domain, which includes operational chemistry and its formidable efficiency (we can add thermodynamics). As the authors themselves make clear, quantum physics offers a vertical explanation, whereas chemistry, with its structural and reactive models, provides a horizontal explanation [WEI 11]. The idea of verticality is interesting, but I prefer to use the concept of thickness. Quantum physics gives access to matter heavily loaded with information, which cannot be limited to the superficial descriptions we so often find, but which are nonetheless useful for representing chemical mechanisms.

Thus, the horizontal explanation resembles a kind of structural GPS which helps the chemist to orient himself in the arrangement of chemical substances and reactions. The reaction between A and B leads to the substances C and D. Horizontality is explained by a simple formula: Water = H_2O. A few grams of water are made up of millions of millions of billions of H_2O molecules. Or better, methane gas = CH_4. A few centiliters of alcohol are explained by the following formula: ethanol = CH_3-CH_2-OH, this time, with a developed formula that makes it possible to predict what the affinity or reactivity of this molecule will be when it is exposed to the presence of reagents. The horizontal explanation is included in the curriculum for chemistry students. For instance, this is the case of the SN2 reaction (bimolecular nucleophilic substitution). Why is it faster with methyl bromide than with methyl chloride? In order to answer this, we must study the bond's stability and examine the solidity that connects the group which detaches itself during the reaction. It turns out that the bromide ion is a "weaker" base than the chloride ion, and that weaker bases detach themselves more easily. As a result, methyl bromide reacts faster. The authors, who sometimes see chemistry's horizontal explanations as a kind of tautology, deliberately chose this example. It is similar to saying that a cheetah moves faster than a camel because its muscles are more powerful and its joints more mobile. Other examples of "tautology" can be found in the formulas of chemical thermodynamics. A reaction is possible because free G enthalpy has a certain value. However, G is calculated by making the reaction happen. Thus, circularity is found in a number of scientific explanations which borrow the principle of horizontality.

Classical chemistry is incredibly efficient. It uses molecular models, which are like Lego games whose support has become unavoidable in chemistry classes. The theoretical arsenal is more elaborate and complex, but it borrows a lot from classical images, with the design of covalent bonds, which constitute the A to Z of chemistry. A covalent bond is based on the sharing of two electrons by two atoms. It is as simple as two persons shaking hands. Covalent bonds are not only essential in chemistry, but also in biology. They are supplemented by other types of forces that make it possible to explain the molecular phenomena of chemistry and life. Ionic forces, hydrophobes, van der Waalls... The notions of "molecular structure" and "chemical bond" were developed well before the quantum era and are of formidable efficiency for carrying out laboratory experiments. However, the structural design of the *bond* does not suffice to describe nature at a quantum level.

These remarks are not a mere detail, as they are involved in two great scientific conjectures, one concerning the status of modern science, and the other, the explanatory role of quantum physics. Chemistry is an effective science that uses formulas, models and theories, but remains inaccessible to the fundamental roots of the things it manipulates. The modern natural sciences can describe phenomena and make use of things without knowing them in depth. The other conjecture is more interesting. It concerns the transition from a quantum description to a classical description. Specialized literature is full of studies drawn from Bohr's correspondences between the classical and quantum worlds. What is less known is that a similar conjecture is delineated in chemistry, with two descriptive fields: a *quantum* field with electronic orbitals and a *classical* field, involving structural chemistry and reactions studied at the laboratory. By combining the two, we meet the molecular model with its atoms, structures and bonds.

Thus, the correspondence between the quantum atom and the world of chemical and/or living substances can be placed at the same level as the correspondence between quantum particles and the classical physical world, with an additional degree of complication. Three levels are required. That of the orbital and quantum communications, that of the molecules made of atoms, and finally, the macroscopic world that we define as specifically classical, which involves laboratory manipulation and an arsenal of theoretical notions, such as free enthalpy, redox potential and PH.

What we must bear in mind is that chemistry differs from classical mechanics in that it studies processes that cannot be interpreted from a mechanistic point of view, that is, they need to be described by notions foreign to Newton's world. There are no forces, but affinities between molecules. Affinity and molecular reactivity phenomena cannot be explained thanks to classical physics but belong to the sphere of quantum physics, which is a science of communications. This is what we will discuss in the following sections.

5.3. Orbitals and waves make bonding improbable

5.3.1. *The electron takes part in the bond, but there is no bond*

A large part of chemistry, the one that studies molecular structures and reactions, is based on the notion of a bond, conceived as the reunion of a pair of electrons that enables atoms to be held together and molecules to be stable. Thus, atoms use one of the peripheral electrons as a sort of glue which makes adhesion to another atom possible, because this also possesses this "electronic glue". This mechanical image still works. It was formulated in the late 19th Century, particularly by Lewis, and it helps us visualize the construction of molecules, thus confirming the atomistic paradigm itself forged at the same time. The methane molecule is a simple example. Its chemical formula is CH_4. The carbon atom contains six electrons, two on the first layer and four on the second layer, which can actually accommodate eight electrons and become saturated. The hydrogen atom has one electron. And so, four H atoms can "lend" one electron each, so that carbon saturates its electronic layer. The mechanism is reciprocal because hydrogen can also saturate its layer with two electrons, one of them being "lent" by the carbon. We nonetheless understand that the mechanistic image is misleading. In the case of the screw and the nut, the two pieces of metal are complementary and remain fixed, while the saturation of electronic layers can only be performed alternately. This is not surprising, since matter is vibration. Besides, the image of saturated layers assumes that the electrons are individualized, but is this really the case?

The ontology of quantum physics is incompatible with the apparent ontology of interatomic bonds produced by electrons ([WEI 11], 4.3). This ontology is said to be apparent because it does not describe the molecule at its most fundamental level. It works as a structural tool for describing molecules from atoms and for imagining bonds as if they were interlocked with electrons. Now, quantum physics tells us that it is impossible to individualize electrons because, even according to the most radical interpretations, these do not exist as individual quantum objects. In addition, these electrons that do not exist are described by a wave function which does not provide them with any position. The electrons are "somewhere" around the nucleus. Electrons "exist", but are nowhere. Last, quantum mechanics tells us that we should not expect to describe electron pairs as pieces of glue located between atomic nuclei. However, the methane molecule exists and methane gas has well-known physical properties, because this substance generates the greenhouse effect that the media so frequently mention. The same applies to benzene (C_6H_6), ethanol (CH_3-CH_2-OH in its semi-developed formula) and millions of molecules whose study is facilitated by apparent ontology. So why cannot we be content with these easily understood formulas?

While chemists do not worry excessively about this question, philosophers consider that behind the evidence dwell questions about nature, time, being and, in this case, the essence of material molecules. In order to solve this "ontological dilemma", one solution would be to admit that while bonding exists, we are still unable to individualize it because quantum mechanics impedes it. In the end, atomic chemistry perfectly describes structures that can be combined, assembled and decomposed during reactions, when the molecules are inside a reactor. It also favors efficient calculations, but reality at its finest level is different. In the classical sense, chemical bonds do not exist, although the bondings are fairly real.

In order to examine the existence of chemical bonds, we will consider the fullerene molecule, which is composed of 60 carbon atoms and whose structure resembles a football. We can place an argon atom in its center and fix it, which would oblige us to admit the presence of 60 interatomic bonds, if such bondings existed. This would suppose a problem, since the argon atom generally forms no bond (such as helium and all the rare gases, whose electronic layer is saturated). In order to explain these phenomena, half of the scientists state that according to one point of view, the connections exist, but that according to another point of view, they might not exist. This constitutes a kind of complementarity in a similar sense to what Bohr suggested about quantum mechanics. If we look at the molecule from one point of view, the connection exists, but according to another point of view, the molecule is no longer there. This vaguely reminds us of quantum uncertainty: if we observe speed precisely, the position of the particle escapes us. These remarks make us realize that the chemical bond is an effective notion in chemistry's know-how, but that at the microscopic and quantum level, the bond does not exist as a mechanism.

5.3.2. *The molecule in question*

Chemistry uses static molecular models because these are effective for describing the reactions and the assembly of atoms. But reality is no longer represented with this reading grid inherited from atomism and Newtonian physics. Molecules are not static entities and representations are often misleading, like that of the water molecule described with two O-H bonds whose angle is calculated exactly at 104.5 degrees. What is more, the notion of molecular form dissolves with quantum mechanics. For some, the molecular form emerges from microphysical interactions and constitutes a relational property, not an intrinsic one. The shape of a molecule is an approximate representation of the relations that connect atoms among each other with the interatomic bonds related to electrodynamic interactions. This form is idealized. The microscopic structure of molecules is not obvious, but its dynamic aspect is irrefutable. If we take the example of water once again, the molecule is not fixed, but it is in perpetual interaction, due to hydrogen's lability, which does not remain stuck with the oxygen and is detached in the form of an H^+ ion, so as to later recompose with the complementary OH^- ion, and constitute the water molecule again. Besides, the average concentration of these ions is described by an easily verifiable macroscopic property, the PH, which can also be detected by one of our five senses, because we can certainly feel the acidity of a lemon juice or a gherkin out of vinegar. Finally, the water molecule makes it possible to make assemblies from a very large number of components, even without having the stability of polymers governed by covalent bonds, such as PVC or the DNA molecule. The water molecule is not a simple component. When it is studied at the molecular level, it shows microscopic properties in large numbers, with a plasticity that can be found in the properties of liquid water which, in addition to its astonishing chemical faculties, is an essential element that makes it possible for living systems to function.

The chemistry universe raises ontological questions which have remained unresolved so far because, while we manage to synthesize millions of molecules, describe their reactions and what they look like, we still do not know what a molecule really is, if it has properties in itself or if these properties emerge when there is a sufficient number of molecules to constitute a macroscopic (or mesoscopic) system. "Chemical matter" is reduced to the properties of individual molecules, but the primordial physical entity imposes a description with the tools of quantum mechanics. This leads to a paradox, since the quantum level dissolves the individual existence of these basic quasi-classical elements, most often represented within space. In short, the individual molecule, conformed by a certain structure, loses its primitive characters at the quantum level at the time that it collectively acquires systemic properties. This is not only the case of water, but also that of all the molecules whose large number generates emergent properties which, most of the time, cannot be apprehended by calculations or structural properties. Ethanol and dimethyl ether share the same chemical formula, but one boils at 78 degrees, the other at 34; one is soluble in water, the other very little; one is toxic, the other is not. The world of chemistry offers astonishing ontological perspectives, especially for understanding how life works. Because even for abiotic molecules, properties are not obvious and cannot be deduced from a simple addition of elementary properties. This is what we call *emergence*. We understand it by studying water, the states of matter, phase changes, solubility and acidity properties, as well as the chemical reactions whose complete progress cannot be elucidated because it lacks the necessary information for describing the unfolding of events at the finest level. This leads chemists to use a simplified substitution model based on a succession of stages (modeling the transition complex, for example, where departure and arrival products coexist). These models, defined as *thin notions*, appear in all chemistry textbooks aimed at students.

5.4. Quantum bond, orbitals and monadological conception of chemistry

5.4.1. *The theory of molecular orbitals in chemistry*

The study of quantum physics enabled us to envision a monadological concept of matter, a substance that undulates, vibrates and whose natural role is to transmit information through processes known as decoherence or the photoelectric effect. Quantum physics describes a very unique spectrum of phenomena and properties concerning electrons, the interactions between charged particles and the EM field, as well as different characters of elementary "particles", spin, isospin, magnetic momentum, kinetics, energy, U(1), SU (2), SU (3) symmetries, quarks, and so on. It is often necessary to incorporate representation involving angles and distances. Representation is useful in chemistry because a bond between atoms supposes an orientation and an arrangement of the electronic "layers".

Thus, quantum chemistry deserves a special analysis because it provides details about the genesis of molecules and of electronic orbitals. The structure of molecules is based on the description of the bonds. Three theories are used frequently. The first one assumes the legacy of the conceptions imagined by Lewis a century ago, with electron sharing. Two other theories derive from quantum mechanics. The first one uses orbital hybridization (*valence bond theory, VB*). In quantum physics, the orbital is a function supposed to describe the probability of the presence of an electron in a small volume of space around the nucleus (in the case of a stationary state), with mathematical precision. Chemists use the concept of an electronic cloud. These clouds can be represented by figures and calculated thanks to Schrödinger's equation applied to the hydrogen atom with the use of three spatial coordinates, one describing dimension (it represents the distance between the electron and the nucleus), while the two others describing angles and are dimensionless.

Solutions are multiple and depend on three quantum numbers that play very specific roles: n, l, m ([RIV 89], Chapters 1 and 2). Let us take a look at the example of an R (r) radial function and a Y (θ, χ) angular function:

$$R\ 2, 1 = (Z/a)^{3/2}\ 1/2\sqrt{6}.\ Zr/a.\ e^{-Zr/2a}$$

$$Y\ 3, +1 = \sqrt{31}/8\sqrt{\pi}.\ \sin\theta\ (5(\cos\theta)^2 - 1)\ e^{i\chi}.$$

Number N defines the spatial expansion (volume) of the electronic cloud, whereas the two other numbers are involved in the orientation of the cloud. N makes it possible to define the electronic layers designated by letters s, p, d, f and g. Hybridization-based theory uses these orbitals to combine them and thus provide the "shape" of the bonds between atoms. Bonding orbitals are designed by combining atomic orbitals located on different levels. Two s electrons will occupy a *sigma* bond orbital. For p, we will have *pi* and for d, we will have a *delta* bonding orbital (which concerns the transition metals such as iron, in particular). The theory based on valence and hybridization overlaps molecular orbitals in order to find, thanks to individual electronic shapes, the classical structures of bonds with their number of shared electrons, and if possible, the angles and the structures of the molecule. The method works for diatomic molecules (O_2, CO, N_2 and NO, for example) and simple molecules such as methane (CH_4), but numerical calculations are quickly exceeded when it is necessary to consider molecules of respectable size.

There is a third theory of chemical bonding, that of molecular orbitals whose principle is to combine the functions of electronic waves by using the principles of quantum physics, which leads to providing representations where, contrary to the *VB* hybrid theory that uniquely considers bonds, the pairs of bonding electrons are not specifically localized, but distributed all around the nuclei.

5.4.2. *Monadology, bonding and non-bonding orbitals*

The theory of orbital-based chemical bonding uses two key concepts for understanding the formation of molecules starting from atoms. Atomic electrons are described by a wave function that represents the shape of an "electronic cloud" and which defines an orbital, later occupied by the electron. When atoms approach (for communicating), the electrons influence each other. Quantum chemistry uses the calculation of a wave function occupied by two electrons "being shared" by two nuclei. The spatial properties of electrons are contained in the orbital wave functions and their respective squares, which indicate the electron density. Let us take two atoms, A and B. Bonding orbitals are simply calculated from atomic orbitals.

Function of the bonding orbital: $\psi_{MO} = \psi_{AO}(A) + \psi_{AO}(B)$

Density of bonding electrons: $A^2 + B^2 + 2AB$

Function of the non-bonding orbital: $\psi_{MO^*} = \psi_{AO}(A) - \psi_{AO}(B)$

Density of non-bonding electrons: $A^2 + B^2 - 2AB$.

Given a homonuclear diatomic molecule with A and B for designating the two atoms (H, for example), the term AB positively intervenes in the bonding molecular orbital (MO), and the "electronic interferences" are constructive. The opposite occurs with the non-bonding molecular orbital (MO*), where the interferences are destructive and can be interpreted in several ways. Everything happens as if the electrons were attracting or repelling each other. In the first case, the electrons are in "phase", whereas in the second, they are in opposite "phase". These images provide the electronic connection with physical meaning rooted in the Newtonian paradigm of forces acting in space. In the new paradigm, the electron participates in a dynamic monadology, involving a two-way communication: reception and expression. In this context, the bonding orbital corresponds to a resonant configuration of the "monadological field", in other words, the communicating interface where expression and reception meet. A bonding orbital describes an interface in which two atoms are in phase and meet in order to "communicate". A non-bonding orbital describes the opposite, that is, two atoms whose electrons "turn their backs" to each other. While this interpretation does not provide a practical application, it nonetheless offers meaning so as to better understand nature from the perspective of an information-centered paradigm. A diatomic molecule is then interpreted as two atoms having entered into an interface in order to "exchange" information, in the form of resonance. Resonance should be pictured as a musical note played continuously, at least as long as the molecule remains stable. Either two atoms bond together and mutually agree, or they fail to get along if they do not bond.

In the H_2 molecule that makes up hydrogen gas, the two bonding electrons have opposite spins and occupy a composite *sigma* 1sσ orbital whose energy level is lower than that of a single orbital that describes the atomic electron and is defined as bonding. It is for this reason that the molecule of dihydrogen is more stable than the atom considered separately. The s atomic orbital makes it possible to form a bonding orbital, 1sσ, and a non-bonding orbital, 1sσ*. Other orbitals are planned, for example, 2π2 and 2π*, and so on. The formation of molecules is determined by a bonding order used in structural chemistry, which can be calculated in quantum chemistry by subtracting the number of bonding electrons from the number of non-bonding electrons, and then dividing by two. If we make the calculation for hydrogen, the bonding order is equal to 1 because there are only two bonding electrons. Dihydrogen H_2 exists! For helium, we have two bonding electrons on the 1sσ orbital and two non-bonding electrons on the 1sσ* orbital. The bonding order equals 0. Conclusion, dihelium, He_2, does not exist!

The structure of the molecule and of "electronic clouds" is crucial. While the musical note gives the image of information transmission between atoms, this note does not propagate in all directions because it is oriented, thanks to the guide of orbital structures. While the monadological link can be interpreted as resonance, it is necessary to specify the orientation of the "monadological clouds" around each nucleus, as well as their extension, indicated by the energetic level of the orbitals calculated with Schrödinger's equation. This formalism determines the configurations that make it possible to describe the orbitals for every atom, s, p, d, f and g, but we still do not understand why this procedure works so well. The rule of structures is certainly intertwined with the rules of communications and the resulting bonds. This is what can be revealed with the descriptions of quantum chemistry: a universe where everything is orientation and angles.

5.5. Molecular quantum communication

The old chemistry used to work with the principle of hooked atoms attached with electrons which served as mechanical parts like screws and nuts in metal constructions. The new chemistry uses other images, resonances, instruments helping to transmit notes and establish bonds. Atoms form bonds using electrons, understood as communicating interfaces. The interpretation of quantum physics in terms of communication and information also applies to quantum chemistry and gives a succinct explanation of the processes involved in interatomic bonds which, we should remember, are at the basis of all molecules as well as the reactions they can produce.

The hypothesis of communicating atoms and molecules will be confirmed with quantum electrodynamics. This theory envisions the exchange of virtual particles that mediate electrical interaction. Material elements, atoms or molecules, repel or attract each other in response to communication processes. Van der Waals "forces" are produced by low intensity electrostatic interactions. They can only be explained thanks to a quantum field theory that foresees the exchange of virtual particles, but these forces are weak and without a measure in common with chemical bonds, because the electrons located on the atomic orbitals do not interfere and remain arranged on their layer. In some configurations, these forces can become powerful. They can be found at the origin of the adhesive properties of the legs of the gecko, a lizard that is able to climb the smooth glass of a building. The virtual particles exchanged by van der Waals "forces" are interpreted as matter communication inaccessible to our experience. We can theoretically deduce them; they are physical and are not calculation tricks. Let us also observe the importance of orientations in these of van der Waalls phenomena.

Matter exchanges information and reacts to this information by arranging itself in a certain way. This phenomenon is defined as bonding. Physicists explain bonding by resorting to a classical notion, *force*. This is sometimes referred to as interaction. Hence, a bond is conceived as the result of communication made with interfaces. It is these communications that take part in the emergence of properties observed by classical science: the three states of matter, gas, liquid, solid; then the emergence of statistical processes as observed in thermodynamics; and finally, these two other "states of matter", which are life and consciousness. Gravity is also conceivable as emergence. It is legitimate to think that these emergences, as for example the liquid phase with its viscosity, are based on a communication mode (*communic-action*), generated by matter components with a certain energy, which – by reaction – will trigger a reorganization of arrangements (molecular agitation) in order or in disorder. Gas is the highest agitation phase, then we have liquid, and finally the solid phase. During each of the phases, particles communicate and exchange information (which is also energy). Absolute zero defines a state of matter in which there is no longer any kind of communication.

Quantum chemistry makes it possible to understand the role of quantum communications in the emergence of a classical world, with its macroscopic characters, as studied by modern science. The descriptions of classical physics are approximations that disguise quantum communications, whose role at the level of large sets of particles may be neglected if we limit ourselves to practice, technique and matter manipulation. Thermodynamics can do without quantum communications for describing energy exchanges and the evolution of systems towards equilibrium. Classical chemistry offers recipes for producing substances from reagents, but circumvents the description of the microscopic and quantum processes involved in these reactions. The description includes angle-geometry used for calculating the orbitals and the energy that describes the power of communications in quantum chemical bonds. Quantum physics places the orientation of matter and information at the center of its descriptions. A new chapter opens, which invites us to understand emergences starting from the communications of matter.

6

Emergences in Matter

6.1. Introduction about emergence

Emergence is a relatively recent concept that went through serious transformation since its introduction in philosophy in the late 19th Century. Its formulation was echoed by the systemic currents of the 1950s, whose founding statement was: "the whole is greater than the sum of the parts." In other words, the parts are not strange to each other, but get together to produce a "composite system", which presents emerging properties. That is, these are properties produced by the collective, and solely by the collective. In fact, the parts taken separately cannot produce the characters, properties and phenomena accomplished by these same parts when they constitute the "whole". From this, an epistemological consequence is derived: the properties of the system cannot be fully explained by studying its parts separately. If we pick up a cell, break it up into elements, and then try to make it work, it is impossible. We may well analyze all the molecules of the cell but we will not find in any of them, even in DNA, the emergent property that we call the "vitality" or "replicability" of the cell. Emergence is a fundamental concept for understanding the living as well as consciousness, which is also an emerging property related to the human brain. It is also a concept that has been applied since the "great era of systemics" to the sets of physical components placed in a non-equilibrium configuration. Snow crystals formed on the window-panes are also emergences produced from a single component, the water molecule. On a completely different scale, the social one, language is an emerging structure resulting from communication between humans. The city is an emerging societal form based on a particular type of communication. The language of birds is also an example of ecosystemic emergence.

Since modern science started analyzing the components of natural things, the question of emergence has become not central but universal. Not only consciousness or life, but all the properties of matter are emergences. Even the most inert materials should be considered as carriers of emerging properties which cannot be explained by separately analyzing their components. This is the case of the crystals produced by certain minerals; it is also the case of phonons which propagate in solids and besides, each phase of matter is a property that cannot be explained at an atomic level. In a solid, the arrangement of atoms is fixed. In a gas, atoms are loosened, whereas in a liquid, atoms modify their arrangement in a reduced volume, as if they were assembled by “fluid glue”. The character of any of these phases can be explained with the properties of components taken separately. We will not find viscosity or solvency in the H_2O molecule.

Language, the arts, cities and every human construction involve subjective access to the understanding of emergence, since each individual is a stakeholder and a conscious actor in the process, even if they are not always the owner builder. The emergence in living systems can be understood from a different angle because we are external factors to the phenomena observed and we can therefore only draw conjectures, hypotheses and explanatory theories through study with the aid of a “technological tin opener”. Some properties of the living are close to us, like the ones developed by animals in their environment, while others are quite enigmatic, as for example the replication of a cell and, particularly, the emergence of life. Striving to understand the living organism with the method of the human sciences would be equivalent to picturing oneself at the place of a cell and experiencing cell life in the company of other cells, or even imagining oneself as a RNA molecule, for example. Finally, emergence in matter represents another scientific challenge, especially if we take the quantum level and the rules of this strange material world into account.

I suggest this statement: *every emergence lies upon two orders of reality: on the one hand, communication between components and, on the other hand, the achievement of a submerged order. Communications are then regulated, and actions coordinated.*

In our times, the order of matter is a challenge that seeks to go beyond a positivist science which contents itself with describing the rules of measured facts. This stake concerns not only physics, but also biology with its explanation of life and, ultimately, the emergence of consciousness. A quick glance shows that the question of emergence in the systemic approach (Bertalanffy) found its origin in the exploration of biology and society as study objects. Research on non-equilibrium systems also makes part of systemics, with the use of quantum theory by Prigogine and his school. Thousands of articles devoted to emergence were published in Anglo-Saxon journals these last decades. The 21st Century saw the emergence of a second quantum revolution with the notions of decoherence, and especially, entanglement. The place of physics is decisive because this science describes the behavior of matter, the substance everything in nature is made of. It is necessary to be cautious and not become too enthusiastic with a quantum mechanics that, alone, does not suffice to explain emergences [BIS 06].

The scientific philosophy of emergence will be as fundamental as evolutionism was during the 20th Century. However, this science will not be able to develop from an approach that connects the physical with the biological sciences. We must differentiate the question of self-organization inherited from non-equilibrium thermodynamics and the question of emergence, which largely surpasses it on many epistemological and ontological points. And this is done by using an ascending and descending method. Physical descriptions are not only useful for explaining living systems, but also the findings derived from the analysis of biological processes can enrich physical theories. This possibility was envisioned and clearly exposed by Paul Davies, physicist and transdisciplinary researcher.

6.2. The central conjecture in emergence theory

The universal and central conjecture of emergence can also be described by a negative statement: the whole cannot be understood by limiting itself to a study of the parts. This conjecture is explained by a large number of systematicians who give a particular but nonetheless pertinent version of the approach, shedding light over the different possibilities that configure it. One of the experts in the field, Peter Corning, has a definite opinion about this question:

> "Rules, or laws, have no causal efficacy; they do not in fact 'generate' anything. They serve merely to describe regularities and consistent relationships in nature. These patterns may be very illuminating and important, but the underlying causal agencies must be separately specified (though often they are not)". (Wikipedia, Emergence)

Corning uses a metaphor that is both illuminating and misleading, suggesting that chess rules cannot explain the unfolding of the game because it is the players who move the pieces of the chessboard. This metaphor is useful for understanding that the system with its emergences has determinations that cannot be reduced to the laws of its components. We understand why this metaphor is misleading. If it described reality, we would have to admit that the pieces and the chessboard are capable of generating the chess players. Nevertheless, this metaphor is illuminating in the measure that it shows a certain heteronomy by which controlling (vertical) instances exert a certain constraint on components, which enables them to behave collectively. Paul Davies suggests that life rests on a hierarchy that regulates the information that flows through organic molecules with top-down causality, from the control center to peripheral information. Going back to the metaphor, the shape of the pieces, their properties and their communications are described by physical laws. The chess game with its rules about moves and its strategy describes how life emerges from molecules. Emergence lies on an efficient instance that organizes imagined life as game of chess with pieces that are "fully engaged". Emergence is not connected to the physical laws of molecules, any more than the shape of the chess pieces can explain the rules and the game that is played. "The system involves more than the rules of the game" (Corning). The systemic conjecture was never solved, but has been widely commented on and explained, particularly by the scientists who adhere to the thesis of strong emergence, by assuming the presence of a causal instance acting upon the components in order to organize them:

> "Although strong emergence is logically possible, it is uncomfortably like magic. How does an irreducible but supervenient downward causal power arise, since by definition it cannot be due to the aggregation of the micro-level potentialities? Such causal powers would be quite unlike anything within our scientific ken. This not only indicates how they will discomfort reasonable forms of materialism. Their mysteriousness will only heighten the traditional worry that emergence entails illegitimately getting something from nothing". (Mark Bedau, Wikipedia, Emergence)

That something which emerges from nothing is reminiscent of Leibniz's famous question: "Why is there something rather than nothing?" Might the question of emergence be in accordance with Heidegger's interrogation of Being, which is as much about Being as it is about time? There is emergence because there is Being. If this hypothesis is true, then emergence could give us access to the understanding of Being. Hence, the idea of a book entitled "Being and Emergence". By studying Heidegger's writings closely, we might find an evocation of emergence understood from the "ontological block" that Dasein implies and which travels through time. Let us not forget that emergence is inseparable from time, to the extent that there is a break between the past and the future at the moment when emergence occurs. A new structure, a system, is formed. Emergence and Time: this duet is as much suited to philosophy as it is to science, especially physics (see Chapters 9 and 10).

6.3. Physics, emergence ontology and communications

Quantum physics teaches us that matter communicates. This essential property leads to a complete re-thinking of the question of systems and emergences. Classical conceptions are based on the interaction of components. However, we still need to clarify what we mean by "interaction". There are several types: mechanical interactions, and also transmission of information which, in relation to the quantum of action, can be defined as "*communic-actions*". Let us briefly recall some considerations (*Emergent Properties*, *Stanford Encyclopedia of Philosophy* (SEP), 2015).

1) Epistemological emergence is the oldest one. It outlines the limits of human knowledge. Emergences cannot be predicted or anticipated from a pre-emerging configuration of the system. Another version postulates the inability to "capture" the properties of the system starting from the notions and concepts used in the physical and dynamic description of the components. Thus, the sensation of the color red cannot be described in terms of the electrical phenomena occuring in neurons. It is pretty trivial. The water molecule does not contain the property of "viscosity".

2) Ontological emergence postulates that the world is made of material components and composites, but that these composites are not always simple aggregates. They are organized according to their degree of complexity in levels, strata and categories of objects. Each level is produced by a set of interactions between components and manifests new qualities. 20th Century physicalism is one of the most widespread conceptions of ontological emergence, especially among materialistic and monistic scientists. Other options are possible, of course. One of them places information at the center of the question, thus enabling an innovative approach that breaks with the conventions inherited from atomism and classical physics, with its somewhat "ambiguous" interactions.

> "As a property of matter, information leads us to a dualism of properties rather than to a dualism of substances. Information is an emerging phenomenon that establishes a difference between the characters expressed by biological systems from the traits that characterize the physical or the chemical components they are made of. Information is neither matter nor energy, but it needs matter in order to incarnate, and energy to communicate". [OCO 15]

Paul Humphreys proposes another hypothesis that conceives emergence as a certain type of "fusion" carried out by the components: "[Emergent properties] result from a fundamental interactive process [fusion] involving constituents with their properties, a process which is nomologically necessary for the properties of the system to emerge. Merging entities lose some of their causal powers and cease to exist as independent elements, while the elements placed in emergence acquire a new causal power". Humphreys suggests that merging is a real physical operation and not a mathematical or logical trick applied to the representation of the system's properties (*Emergent Properties, SEP*). This conception is rather original because it presupposes the disappearance of individual properties and their replacement by the collective properties that the emergent system expresses. This hypothesis is tangible at least on one point. The individual behavior of the components capable of disturbance (by generating noise, especially) and of becoming opposed to the coherence of the totality must be suppressed, or controlled by making them inoperative, silent.

3) The majority of systematicians conceive emergence as a process that essentially involves specific relationships between the components described by the physical sciences and these same components in the way that they behave in complex systems. This vision, dating from the previous century, has been disputed. Contemporary physics teaches us that emergence is not only specific to complex systems, but also characterizes the properties of non-organic systems, and not only just dissipative structures. Condensed matter, as any liquid or solid, is an "emergent" composite material made from the most basic constituents: not only from atoms

in the modern sense of the term, but also information, fields, entangled quantum systems, and so on.

If the physics of the 21st Century offers us new elements for understanding nature, it is with the idea of emergence, which is no longer reserved for complex sciences, and also concerns ordinary material phenomena. Robert Laughlin, winner of the 1998 Nobel Prize in Physics for the discovery of the quantum Hall effect, is convinced that many properties of matter studied at the macroscopic level (crystals, superconductivity, ferromagnetism, phonons, Hall effect, etc.) can only be understood by principles applied to the collective of components, and not in terms of the fundamental physics that describes elementary components. Laughlin suggests replacing reductionism with emergence, both epistemologically and ontologically. From this author, we can read these very enlightening observations:

> "The generic low-energy properties of crystalline states are determined by a higher organizing principle and nothing else [...] The consensual belief that the renormalization of the universe is based on a theory of the whole relying on the microphysical order rather than on emergence is nothing but an article of unfalsifiable faith".

Then, regarding the Hall and Josephson effects:

> "None of these phenomena can be deduced from the microphysical level, and both are transcendent, in the sense that they would still be authentic even if the theory of the whole was changed. Thus, the existence of these phenomena is decisive in the measure that it shows that at least, at a certain fundamental level, the descriptions contained in the theory of the whole are not relevant".

To sum up, emergence concerns matter and all its macroscopic properties found in solids, liquids and other more or less exotic phases. A new page is written in the field of physics. Emergence will be the challenge of the 21st Century.

4) Communicational emergence relies on two orders of reality: coherent communications between the elements and the coordinated reactions of the elements with an informational order produced by the causal instances of emergence. Each of these instances comprises two categories: horizontality or verticality (hierarchy). Each component makes a kind of experience in a field of communication and action; it has a memory that enables a type of "knowledge" acquired over the other components. Emergence occurs because components are progressively determined as part of a whole, once they have exchanged a sufficient amount of information. Quantum physics tells us that fundamental matter communicates. We still have to understand how macroscopic matter emerges. And if information is the central notion, how do we go from disordered information to an emerging order with its forms, characters, properties and functions observed by common sense and scientific experience?

The philosophy of emergence and communication differs from epistemological emergence, which refuses to get to the bottom of things, and also differentiates itself from physicalist emergence, which simplifies things. I suggest the idea of an auto-built submerged core that regulates communications, which become more efficient and coherent. Another "core" regulates the actions that are translated on the macroscopic scale by observable properties or, if applicable, by physiological or mental functions. The study of matter makes it possible to collect all the clues supporting the unprecedented thesis of communicational emergence. With the aid of signs, it also helps us to reveal the existence of immersed cores: *there is no emergence without submergence.*

6.4. Quantum and information in the emergences of condensed matter

6.4.1. *Material emergences and physical sciences*

Emergences can be of several types and depend on the things studied and the methods used for describing them. The microphysical components can just as well be described using classical variables, as in thermodynamics, or with quantum formalisms that may also include orbitals and entanglement. According to their type, emergences can be natural, such as the three phases of matter and, specifically, the properties of liquids: solution, viscosity and capillarity; or of solids: phonons, conductivity, magnetism and crystals. Physics studies artificial emergences, that is, produced by materials that require the intervention of human hands and that were placed under special conditions. For example, this is the case of liquid helium or liquid crystals.

The physics of condensed matter studies emergences with three types of approaches: (1) classical, resulting from the mechanical and statistical descriptions of thermodynamics, with perfect gases or with dissipative structures, for instance; (2) by combining a classical approach with a quantum description, as for example in coherent fields and Fermi liquids. We could also add the theoretical emergences that are the three quantized fields (electrodynamics, weak and strong). The quantum character is involved in notions that are specific to this physics: orbitals, spins, fermions and bosons. The classical approach is introduced by Landau's theories and statistical tools, such as density matrices. (3) Finally, certain methods resort to quantum properties, such as entanglement, and new mathematical tools, such as tensor networks.

For decades, quantum physics explained a large number of phenomena, involving atoms, particles and radiation. Now, it is possible to unify quantum information and matter under a certain modality (Wen). The "new quantum physics" unifies information and emergence by studying quantum phase transitions in condensed matter. Emergence then becomes a question about how information is organized and regarding the precise role of communications, if necessary including considerations about the hidden order of information, memory and the possible setting or achievement of elementary communications.

6.4.2. *New horizons in the physics of condensed matter*

A new scientific adventure has started in the last one or two decades, with strange quantum material being interpreted in the context of information science and entanglement being taken into account. This turning point was presented in a book accessible on the prepublication site arXiv; it is signed by one of the best known specialists in condensed media physics, Xiao-Gang Wen, assisted by three collaborators. Its title is *Quantum Information Meets Quantum Matter* and its subtitle, *From Quantum Entanglement to Topological Transitions in Multiparticle Systems*. In its preface, Wen predicted a golden age for research in the context of quantum information and the technologies flowing from it [ZEN 16]. Quantum research will make it possible to construct new notions and formal tools so as to understand the phase transition phenomena observed in the multiple forms of condensed matter and many other things. We traditionally know the three phases, gas, liquid and solid, but according to Wen, there might be hundreds of phases if we consider imaginary experiments by combining certain elements arranged in a certain way and with the adequate temperature. The case of superfluid helium is only an example of what a special phase of matter may represent. Many of these exotic phases use low temperatures and/or magnetic fields. This detail is not trivial, because temperature plays a role on noise and communications (see Chapter 10 for this test), whereas magnetism seems to regulate communications while determining orientations in matter.

Wen announced a paradigm shift in Kuhn's sense. For decades, scientists were convinced that the phases of matter obeyed Landau's theoretical formulation based on an order "crossed" by breaks in symmetry. Now, a new order of matter was discovered by studying electrons that form a two-dimensional layer that can generate quasi-particles. This phenomenon, known as the quantum and fractional Hall effect, was discovered in 1989 and accurately described. However, the microphysical origin of the phenomenon remains enigmatic. Hall's quantum effect is one of many unsolved issues of contemporary physics. This effect presupposes the existence of a type of transition that escapes the theoretical framework of Landau and which heralds topological order in matter as the new paradigm.

According to Wen, the traditional physics of condensed matter considers two categories of states formed by large sets of components. On the one hand, the states produced by "coherent fields" and, on the other hand, the states obtained by filling quantum orbitals, as in the case of Fermi liquids. These two categories fail to encompass the topological order and consequently lead to the question: are we capable of understanding what escapes (or fails to) the traditional order of matter and properly "capture" the new topological order? It is this question that Wen tries to answer. Quantum entanglement, foreign to classical physics, serves as a guiding principle for explaining these new phenomena which arise from matter, and in which a topological order can be read.

The works introduced by Wen describe emergence from an unparalleled angle. Tools that were not used in 20th Century physics are now included, particularly quantum entanglement, which might be at the root of the topological order in the ordered phases of matter. We might also assume the hypothesis of a connection between quantum information and condensed matter. The notion of topology in matter thus might constitute an indicator that reveals submerged and submerging order in matter.

6.4.3. *Matter and information according to Xiao-Gang Wen*

In the last chapter of their book, Wen and his collaborators offer a brief history of physics and then a clarifying explanation regarding the current quantum information revolution. The central thesis that works as a guiding thread of the book is that information unifies matter. This sibylline formula means that matter is no longer understood as it was a few decades ago, centered in the classical notions of particles, atoms, objects, forces and fields. Components themselves are no longer sufficient to determine quantum matter and it is necessary to use the notion of quantum information. In fact, the schism between classical mechanics and quantum physics has deepened. Classical categories do not apply to the quantum world. The classical world is not conceived as an addition of small quantum things. Between these two worlds an emergence is delineated, with a weaving of elementary quantum and microscopic information in order to shape the macroscopic classical universe of material objects, forces, communication fields and arrangements in the cosmos. Radical consequences must be drawn. For example, the notion of elementary forces between particles no longer has any legitimacy.

Leaving thermodynamics aside, four revolutions may have played a key role in modern physics: (1) Newton's mechanics, with masses described as material points moving on a fixed scene; (2) Maxwell's electromagnetism, with fields moving on a fixed scene; (3) the revolution of relativity radically changed the conception of the scene with two representations of space-time, one for radiation, and the other for the arrangement of bodies in the cosmos. The scene of gravity was no longer fixed; (4) The last revolution concerns the minute details of quantum matter, which appeared under a strange form. For some, it was both wave and corpuscle, while other thought that it was neither a wave nor a corpuscle. The quantum revolution included two phases: the first one was prompted by the founding fathers and their research during the 1920s. The second phase is very recent, and it involves quantum information and two new and incomprehensible notions for classical science, entanglement and the qbit (without forgetting decoherence). For Newton, matter was force and inertia; for Einstein, matter was equivalent to geometry; for the fathers of quantum physics, matter was wave and corpuscle; for new physicists of the 21st Century, matter is equivalent, even identical, to quantum information. Wen suggested an unprecedented unification between electronic matter and electromagnetic radiation. Rays of light would then be seen as large-scale fluctuations of quantum entanglement, whereas electrons would result from a deficiency in entanglement (which would thus engender informational singularities).

Then, the contemporary epistemological break might be indicated by a mere additional letter. In the 1980s, Wheeler suggested an equivalence between matter and information, with the famous "*it from bit*" formula, which means "things derive from digital information". The new proposal is stated with the formula "*it from qbit*", which means that things are composed of basic quantum information. When things are studied by physics, they are described by several equations: Maxwell for electromagnetic rays, Dirac for electrons and fermions, Yang–Mills for hadrons and gluons of the strong force then the particles of the weak force, lepton, W/Z boson. Wen later proposed an astonishing hypothesis. All these equations might have a common denominator, quantum information and entanglement, in other words, the order of qbits in the universe. This kind of proposal is likely to disturb the understanding of things. *All scientific revolutions result from the discovery of a hidden unity in the order of phenomena, gravitation and inertia for Newton, electricity and magnetism for Maxwell, matter and geometry for Einstein, matter and wave for de Broglie and the fathers of quantum mechanics. The next revolution could result from a unification of matter and quantum information.* However, this unification does not mean that everything becomes simple. The correct formula really is: everything is complicated! This is what we prove with the quantum Hall effect based on entanglement, as well as the hundreds of exotic states of matter that tensor networks, and thus the topology, "capture".

The scientific revolution of the 21st Century put an end to two hundred years of reductionism. We can no longer understand things by indefinitely dividing them into more and more elementary parts. For example, phonons, which are quanta of sound waves propagating in solids, cannot be reduced to atoms and are conceived as collective and coherent movements of these atoms. They are (mesoscopic or macroscopic) emergences, not based on the interactions between components endowed with properties as in old systemics, but rely on complicated networks that carry quantum information. Complex topological orders are at the source of the phases of matter, with the properties observed. Quantum order determines the formal richness of our material world.

Wen described the new physics with highly technical considerations on quantum liquids, string networks and the astonishing idea of considering electrons no longer as particles, but as the collective and coordinated behavior of "something" that might be related to quantum order. In addition, large-scale qbit entanglement might unify electrons and light rays. We should not forget the settings revealed by the gauges and other invariants that might not be prime elements, but the consequence of the organization of quantum information (related to submerged structures?). Gravity will not escape this turning point linked to tensor networks. We may wonder whether this notion will not play a universal role, like the material point in modern mechanics, or the concept of field in 19th Century physics. Networks and emergence might be a couple with a bright future!

6.5. Tensor networks

6.5.1. *The tools of statistical physics*

The description of quantum systems with N particles uses specific tools for building theoretical models and also for the calculations made by computers for empirical purposes. These systems can be gases or phases of condensed matter, the most common of which are liquids and solids, but whose number is coded by hundreds with strange phases such as superfluid helium or liquid fermions, not to mention liquid crystals, which became famous for the general public. Every science is transformed with the use of central notions, accompanied by mathematical tools. Statistical physics does not escape this rule. The first statistical descriptions were derived from the formulation of entropy by Boltzmann and Gibbs, with the famous S = k.log (W) formula. In the course of the 20th Century, other tools emerged, especially for noting the results of quantum mechanics. For example, von Neumann's density matrix describes a mixture of states at the quantum level, and has also been useful for describing processes studied by the thermodynamics of non-equilibrium. Quantum mechanics and the study of condensed matter phases have enriched statistical descriptions; Landau's work, bosons and fermions statistics, Fermi's gas and liquid, and this new tool, the network of tensors, whose use in statistical physics is extremely recent, barely a decade old.

Tensor networks have been known for a long time and have been used since the 1980s for geometrizing phenomena produced by biological systems, particularly for modeling cerebellum functions. This reveals a similarity between the description of N-particle quantum systems and the dynamics of neural networks. However, we must be cautious, bearing in mind that the models used in neuroscience still belong to the classical world. These use real numbers or functions (as, in general, relativity tensors do), while the tensor networks used in quantum statistical physics are constructed from complex functions (using the i imaginary number, whose square equals -1). The formal similarity between neural and quantum networks may reveal hidden connivances in nature, deeper than the misleading similarity between Boltzmann's entropy and Shannon's formula. As Tonnelat so clearly explained it, a formal analogy does not mean an analogy in nature. Similarities are sometimes contingent, reflecting nomological coincidences, and sometimes meaningful ones, revealing ontological correlations in the principles that govern things.

The use of tensor networks in quantum physics of condensed matter raises many questions. What is the physical sense associated with these mathematical tools? What do these tensors say about our world and about matter? In recent years, tensor networks have not only been widely used for describing condensed matter, but also in quantum cosmology. As a common denominator, this includes the structure of information present not only in quantum matter, but also in the cosmos, via the holographic principle that governs the AdS/CFT duality, for which physicists are still struggling to find a physical meaning. The N-particle quantum systems described with tensor networks are also enigmatic. What do these tensors reveal about nature (*physis*) or about matter (*hylé*)?

6.5.2. *The complicated invisible behind the visible*

Jean Perrin (the father of material atoms) used a clever formula to illustrate the guiding principle of his works: it is necessary to explain the complicated visible by the simple invisible. This precept governed reductionist science for more than a century. The simple invisible, we understand, designates atoms and the elementary constituents that are the particles. Now, quantum mechanics, and more specifically, tensor networks teach us that behind a relatively complex visible, which might be a liquid or even an exotic phase, there might be an extremely complicated invisible. The state vector of the system is described as a network of interconnected tensors. We can intuitively imagine pieces of a Lego game becoming united thanks to mathematical glue, in this case, quantum entanglement. In order to describe the place of the tensors, Orus suggested a comparison with the biological world. The tensor network would be equivalent to DNA making it possible to construct the state vector. In short, if biological DNA with its genes contains the necessary "information" for generating an organism, then the tensor network is like quantum DNA, enabling us to infer the state vector from a multiparticulate system [ORU 14].

Tensors → | Ψ>

DNA → Organism

On my behalf, I suggest a broader comparison that considers three levels: quantum encoding with tensors, quantum transcription with the state vector and an objective translation with the properties of condensed matter observed by the experimenter. In biology, these three elements correspond to the network of genes contained in the DNA, the network of transcribed RNAs, and finally the complete and extended figure with the translated proteins.

Tensors → | Ψ>→ Observed matter

DNA → RNA → Proteins and organism

According to Orus, tensor networks would then constitute the natural language for talking about condensed matter. They might offer more details than those descriptions which have become common, like that of spin liquid. A new language must extend itself so as to cover quantum physics in its entirety; so this is like a second quantum revolution that has begun, including new intuitions about this strange matter. Now, we understand differently how a macroscopic physical reality might emerge from a submerged encoding. The reality in which we live might be encoded at the quantum level and tensor networks could be the instruments making it possible to formalize this encoding, and this including of other fundamental properties, such as plasticity. The general epistemological diagram puts several notions in contact: networks, tensors, state vectors, encoding, observable expression, and also decoding, with, in its center, the key question about material emergence and the properties that result from the coordination of elementary processes.

In short, quantum encoding positions itself as a causal instance and what is observed could be conceived as extended forms resulting quantum decoding! What emerges is thus conceived as a kind of collective decoding, accomplished by the different parts of a system. Finally, dialectics between encoding and decoding can be envisioned, linking emergence to the physical experience of the parties and the communications generated by sharing information.

Another important result concerns morphogenesis. Orus pointed out that entanglement determines geometry. Curvatures and geometries emerge from an entanglement diagram encoded by tensor networks. This principle also concerns the gravity and geometric layout of the extension. The topological order contained in networks, with its encoded entanglement relations, unfolds as geometric order in an extension. This principle will probably remind philosophers of double matter, as theorized by Plotinus in the *Enneads*, as well as sensible intelligible duality in Plato. A scientific philosophy of matter is delineated, thanks to quantum research. The natural states of N-particle systems are better represented with tensor networks than with the addition of Hilbert spaces, commonly used in basic quantum physics. These spaces have to have fairly high dimensions to have physical meaning. Taking entanglement into consideration significantly reduces the size of the system's display space endowed with moderate energy. The actual states of the system can only be described in a tiny corner of Hilbert's space. This "corner" emerges with networks of tensors that perform a kind of sorting among the infinity of possible states (Orus). In the end, Hilbert's space turned out to be a mathematical illusion born of quantum conventions and detached from reality. States produced by entanglement and formalized by interconnected tensors stick better to reality.

Quantum entanglement is also involved in entropy calculations. It makes it possible to sort out the immense possibilities for building states. This sorting occurs at the level of entanglement entropy calculated for an A sub-region, included in a B region. Entropy is then measured by the area that separates regions, not by volume. This property presents a strange similarity with the calculation of the entropy of a cosmological system in which A is inside a black hole and B outside it. However, let us remain cautious because nomological coincidences do not necessarily translate similarities in nature or in principle. Moreover, the meaning of entropy is more of a conjecture rather than evidence. We still have the idea of sorting out, which, if related to a physical property, matter or the human mind, reveals the universal process of suppression, working in nature and in consciousness and present in the ontology of time (see Chapters 9 and 10).

Tensor networks reflect the real topology of quantum information. This notion of topology is rich in meaning. It indicates how an element relates to its surrounding elements. Things are contiguous, not “alien” to each other. Their relation is specified by the graphs linking the elements, be them pieces of information or objects. The situation is completely different in a geometry used for describing extension. Geometric points in space are indifferent to one another. They only have a formal relationship thanks to the positioning that an outside observer assigns to them using the three Cartesian coordinates. It is the observer who establishes spatial relationships between things arranged, whereas topology described by tensor networks reflects the internal organization of quantum information in matter.

Orus represented the topological order produced by tensor networks with graphs, by evoking a similarity of principle with Feynman’s diagrams in order to simplify the quantum processes in electrodynamics. Besides, these diagrams made it possible to simplify calculations, what was again the case of contracted tensor networks, whose effect was described as “magical” by Orus. An N-rank tensor contains a number of coefficients coded at p^N. These coefficients are not independent and reflect the structure produced by the network’s contraction. Hence, this idea of “structural glue” unites the different pieces of the system's "quantum DNA". For Orus, this property is not only mathematical, but also possesses a deep physical meaning because it reveals the internal structure and topology of an N-particle system, thanks to the interconnection present in the elements of the Ψ state vector.

Summarizing, quantum entanglement selects the possible combinations in the different blocks of quantum information, contained in a particle system, in such a way that this system expresses coherence. Network entanglement translates the presence of a submerged structure that engenders an objectified and emergent structure from its components. If an N-particle quantum system creates emergences, it is thanks to the communication between microscopic components, accompanied by the organization of information. Entanglement and the tensor network describe this process whose dynamic cannot be captured by Hilbert's spaces.

In order to observe an entangled system, it is necessary to prepare it by making two "particles" interact. Entanglement takes place after communication. This suggests the hypothesis of an emergence conceived from the quantum point of view and based not only on elementary communications, but also on a submerged order that builds itself by playing on the entanglements. Emergence = communications + entanglements. This brief equation serves as an epilogue to this chapter in which we pointed out the importance of condensed matter in the understanding of physical emergences.

7

Communication and Emergence Fields

7.1. Communication fields

7.1.1. *Quantum communications*

Quantum information makes it possible to conceive emergence that envisions two universal processes. On the one hand, the pole of communication through quantum interfaces, and on the other hand, the deep dynamic of quantum information, deriving from the description of basic states, state vectors and entangled structures. Emergence relies on a sequence of processes that includes the extraction of quantum information, its reception and finally, information order (transformation), which supposedly influences the informed system.

A new philosophy of nature is outlined and can be defined as “communicational emergence”. We can now risk a universal statement: *without “communication and information”, there cannot be emergence*. Because emergence is fundamentally a transformation, and for a transformation to occur, things must be “informed”. Communications are established between the elements of a composite system, atoms in condensed matter, molecules or cells in biological systems, or between a system and its environment, which is the case for specimens in every animal species.

This statement imposes a condition for emergence to occur. Components must be equipped with adjusted interfaces for receiving information, and then a system capable of reading, interpreting, contextualizing and executing a set of reactions, which can be of two types: either a change in the arrangement (position and orientation) or a reaction in the form of signal transmission, and therefore, in communication. This process is dialectical, in fact, and two conditions must be respected: the receptive capability of signals through interfaces, with a scenic setting, and a certain "semantic coherence" in the pieces of information exchanged. Setting was studied in my previous essay, describing the world's adjusted scene that enables material actors to communicate [DUG 17]. However, it is not only the scene that is regulated. The components of an integrated system exchange signals, involving expression and reception through interfaces. Coherence is already evident at the level of the matter that can be questioned. Do physical gauges reflect the existence of settings and coherence in the most basic signals? Can the world be conceived as an ordered "whole" having emerged from a selection process resulting in gauges and bosons?

The hypothesis of "informational Darwinism" (or "semantic Darwinism", if applicable) states that certain signals are selected to become invariants, thus enabling the components and the actors of an emerging system to organize transmission, reception and coordinated actions as well as an "interpretation" of these signals. A signal that is functional continues to be used in the information milieu. A signal that is no longer used ends up disappearing from the communication field. By observing the different reigns of being, matter, life and society, we can detect the settings specific to each communication system. By examining the molecules of the living or the structure of human languages, we will find invariants or ubiquitous messengers. This happens with matter, as well. The understanding of emergence requires the study of communications. This observation is also valid for studying morphogenesis, a particular case of emergence. And in both cases, the global configuration requires the use of some concepts adapted to this type of situation, especially the notion of a "field".

7.1.2. *Morphogenetic field or communication field?*

At the beginning of the 20th Century, the notion of a field, derived from physics, was used for trying to explain the development of organisms. The notion of a morphogenetic field was decisive at the time, apart from being enriched by discoveries such as heredity, the hypothesis of genes and chromosomes. The notion of genes is inscribed in the mechanistic paradigm and, more precisely, in its atomistic version. Later, during the post-war period, biologists left the concept of the field aside and preferred to bet on the paradigm of genes and mutations. The discovery of DNA made it possible to forge the notion of genetic programs, thus confirming the power of the genetic paradigm. In the 1990s, the morphogenetic field was again considered no longer opposed to gene mechanisms, but complementing it. One of the promising hypotheses was that DNA produces components that generate a morphogenetic field which, in turn, influences the genome expression: "The products of genes must first interact in order to generate efficient morphogenetic fields. A change in these fields influences the way in which organisms develop." [GIL 96]

The study of morphogenetic processes paves the way for understanding emergence. According to Scott Gilbert, it makes it possible to renew research about evolution and, particularly, about what escapes the contemporary (synthetic) standard model: the divergence of phylums, with their branches analyzed by taxonomists and sorted out in class, order, family, genus, species. I personally conclude that evolution is a complex process for which the concept of emergence is relevant, and it complements the concept of mutation. These few observations regarding biology introduce a topic that can be articulated around three concepts: communication, fields and emergence.

However, it is necessary to hone our description. The notion of a field is derived from physics, and possesses evident theoretical interest as well as descriptive efficiency, but the concept might be misleading if we essentialize or materialize it, and it should remain a nomological tool above all. If there is a field, it is a communication medium. It is the entities that capture the signals which react as if they were influenced by the field.

In the living world, there is no morphogenetic field *stricto sensu*. Cells communicate, share their experiences, and experiment in the systemic environment. If there is a field, it is an organized environment for the propagation of molecular signals, a medium that is philosophically considered as a field of form, a field of surface and interfaces. The morphogenetic process is determined by the interpretation and response of the organizing seeds which carry the system's memory and the content (which will be assimilated to compressed information as this relates to experiential time, see Chapter 10). The morphogenetic field is only a descriptive trick. Morphogenesis occurs thanks to the interpretation of signals and content decompression (gene expression for biologists). The following considerations attempt to clarify these general ideas about communication fields. Not without revolutionalizing the notion of a field, so dear to contemporary science.

Summarizing, *the field of communications is of the form/space kind, while the operational, cognitive and efficient functioning of the system's components is of the content/time kind.*

7.2. Are communication fields structured by quantum matter?

7.2.1. *The field as a physical concept*

The concept of the field is decisive in physics. It helps us to describe the propagation of influences whose effects are often mechanical, for example, the electric field that produces a force if placed near a charge. The field also determines the shape of something extended. Most physicists think that fields produce forces. However, what the new paradigm suggests is that the physical content of the field should be imagined together with communications involving transmission, reception, expressions and reactions. A field is generated by source and influence "targets". It is the reaction of the targets that reveals the existence of the field with its signals, excitations and bosons. In the case of the electric field, the targets are also sources, the same as in the Newtonian field of gravity, whose sources are masses which in return receive the influence of the field. Most often, a field designates something different from a material object or a substance. It is an extension, a medium, a portion of space where every point is assigned with a mathematical quantity that can take several forms: scalar, vector, tensor and spinor. A field is not observed directly, but by the physical effect that a target undergoes. We do not see the gravity field, but we observe the stone falling on the ground.

An electric charge arranged around a nucleus captures the "real photons" of the EM field and changes its shape. In the context of a quantum description, a charge also captures the "virtual photons" of the electric field produced by another charge and then interacts with a force, which is either attractive or repulsive. The result of this influence is measured with classical variables such as force, speed or rotation. The field is designed as a scene that separates components and enables them to communicate and, if necessary, to react. "*Communic-action*" and *reaction*. While communications are necessary (but not sufficient) for emergence processes, they are also decisive for keeping the order of an emerging structure. It is by analyzing the description of the fundamental things that have emerged that we can get to know the physical realities captured by the fields.

7.2.2. *The Lagrangian and the symmetries, main access door for studying fields as emergences*

Classical mechanics uses two tools for studying how a physical system will evolve over time. The first one uses the Hamiltonian, equal to the sum of two energies, one kinetic, T or Ev, and the other potential, V or Ep. This gives H = T + V. The other tool is the Lagrangian, equal to the difference between kinetic and potential energies. L = T − V. The Lagrangian calculates the integral of action over a path. The principle of least action (also known as the principle of stationary action) states that a system follows the path that minimizes the action along that trajectory. This principle puzzled physicists. What can be the physical meaning of this quantity equal to the difference between two energies? Physicists could not find a definitive answer. Leonard Susskind imagined a "particle" of the system with supernatural powers to feel all the trajectories and calculate the one that would keep the action to a minimum. If this interpretation has a physical meaning, then the "particle" seems to communicate with nature and behaves as if it were remote controlled by a "field" communicating the "good information" to it.

Hamiltonian mechanics is equivalent to Lagrangian mechanics by means of a mathematical operation. Nevertheless, the Lagrangian is more universal and applies to other formulations, particularly quantum electrodynamics. The principle of stationary action is no longer applied to trajectories, but to possible transitions involving particles. Feynman conceived an integral of the paths (in the sense of processes, not trajectories) for applying the Lagrangian to quantum processes. In the Lagrangian, Susskind perceived a universal way to concisely package all the information about a system. He rejoined Feynman in considering that the effectiveness of the principle of stationary action in rational mechanics assigns a quantum origin to classical physics. The other universal principle is the conservation of energy. Let us now suppose that quantum information is even more fundamental. Energy conservation is then conceived as information conservation. And stationary action is also interpreted in the sense of information exchanged from one physical "place" to another, taking place in a conservative system. The place does not refer to a spatiotemporal assignment but designates a more global order of matter, comprising the realities expressed and a printed or submerged "meta-physical" domain.

The principle of stationary action defines how information is transferred and applies to the formulation of dynamic exchanges occurring in (or with) a field. The majority of theories about the field are based on symmetries designed according to the principle of invariance. Certain symmetries are global and others are local; the description of the state of a system (and its evolution) remains unchanged if we carry out a transformation framed by the gauge. Gravitation also has global symmetry. If we perform an identical translation of two bodies in the cosmos, the "force" of gravity generated by the "attraction" of these two bodies remains the same. In electromagnetism, the physical content is given by the electric and magnetic fields, respectively calculated from the electric and vector potentials. A transformation of the two potentials by a function of space and time does not change the physical content of the field.

The Lagrangian also has symmetric properties. If we swap the q coordinate system with an s system in such a way that $q = q(s,t)$, the description of the system's dynamics remains unchanged. Another transformation leaves the Lagrangian unchanged, and this fact is rarely pointed out. If I add to T and V the same amount of Λ, the Lagrangian remains unchanged because $T + \Lambda - (V + \Lambda) = T - V$. I ignore what this invariance may mean but it deserves to be underlined all the same. A symmetry (or an invariance) in a physical description may mean several things. It must indicate only one property of the mathematical formulas used. Invariances are sometimes considered as redundancies, for example, in the formulation of electromagnetism, where the potential vector may adopt an infinite number of values without modifications in field equations. Carlo Rovelli interprets gauges as "formal mechanical pieces" (*handles*) that make it possible to complement physical observations while describing the interrelation between the different components of the physical universe. *In what concerns my hypothesis, I interpret symmetries and gauges as the universal principles that make it possible for material instances to communicate.* It is not surprising to find these symmetries in fields which are at the same time environments where signals and other types of information can flow.

7.3. Is there a relation between quantified fields and emergences?

Quantum matter is described by three theories of the field. These correspond to the three categories of fundamental interaction, often designated as forces, electromagnetic, weak and strong. These three theories use Lagrangian density and have a common structure with four elements: (1) "Fermions", which in quantum electrodynamics are charged fermions (electrons, protons); for a weak interaction, it is necessary to add neutrons, and in the case of a strong interaction, the fundamental constituents are the quarks; (2) Fermions are described by spinor fields for electromagnetism, which describes the fundamental property of a charged fermion, the spin. For a weak interaction, it is isospin, and for a strong interaction, it is the color of quarks; (3) The third element is the gauge, which determines the symmetry with a transformation group that leaves the Lagrangian invariant, respectively U(1), SU(2) and SU(3); (4) Finally, bosons (or "vectors" of the interaction) generated by gauges and represented by tensors. These are the photon, the three weak W^{+}, W^{-}, Z bosons, and finally, the eight gluons.

Electrons → spins → U(1) → photon

Protons neutrons → isospin → SU(2) → W^{+}, W^{-}, Z

Quarks → colors → SU(3) → 8 gluons

These three theories have been the object of various interpretations, the most radical ones denying the physical existence of fields and even of particles. In my hypothesis, the Lagrangian describes that information circulates so well that the notions of fields and particles arise from a primordial reality. It is information that goes first, and depending on the point of view adopted so as to "see" or to "experience" a material thing, it is the particle or the field that will be relevant. This is particularly relevant in quantum matter. Fermions present themselves as information structures that can communicate with three languages whose rules are strict. And what are these three languages?

Electromagnetism describes a language between the objective forms equipped with a voluminal extension (atoms with their corresponding electrons) and, if necessary, forms going through the spatial extent. Photons are similar to musical notes working in the transmission of phenomena, of visible phenomena, particularly. The Lagrangian formulation of quantum electrodynamics provides an additional indicator about this communicating matter. It uses two operators, creation and annihilation, which are adjoined to each other. And we can interpret these (like the operators generating observables in basic quantum theory) as the two sides of a piece of metal with hollows and bumps. Creations and annihilations are the two sides of a quantum reality that unfolds towards us and retreats to its own "world". Besides, there is symmetry between transmission and reception. We find it in the complex world. Nature is regulated for emitted signals to be adjusted so that they can be received. The interpretation of weak bosons does not flow from a source, especially given the fact that bosons have mass and must therefore be placed, in principle, in the context of a physics of arrangements. There are still the enigmatic gluons. What do these gluons communicate? The question had to be asked.

Let us go back to quantum electrodynamics, which describes how "information" and communication are arranged between charges thanks to the spin. The field of the fermions presents itself with the attributes of the arrangement field in the measure that it is the embodied part of the information that is taken into consideration. Bosons are described as excitations running through a communication field. The gauge field is used for coupling the two fields. This interpretation appears clearly if we analyze the terms of the Lagrangian density formulated in the context of quantum electrodynamics. For the electroweak model and the strong model (quantum chromodynamics), a similar interpretation is possible.

$$\mathcal{L} = -\hbar c \bar{\psi} \left(\gamma_\mu \frac{\partial}{\partial x_\mu} + \frac{mc}{\hbar} \right) \psi - \frac{1}{4} F_{\mu\nu} F_{\mu\nu} - ie\bar{\psi}\gamma_\mu A_\mu \psi$$

The L density of the QED Lagrangian includes a first term that involves an m mass, two fundamental constants, h and c, as well as the fermions' creation and annihilation operators, ψ and ψ-bar. This term is interpreted as a field of matter or rather as embodied information. The second term involves tensor F, which is that of electromagnetism. It then describes the propagation of the bosons and it is a communication field. The last term combines fermion creation and annihilation with electric charge e and electromagnetic potential A. It is a gauge field that is interpreted as the description of (interfacial) communication between charged quantum matter and the electromagnetic field. In quantum chromodynamics, the tensors describe gluons, mediators of communication between quarks. A diagram summarizes this idea:

Arranged fermions → Field and gauge, interface → Boson field

Relativistic cosmology also uses symmetries and tensors. It encourages us to interpret our local cosmos as an arranged structure that has emerged. In Einstein's formulation, matter is not introduced as a spinor field, but as a stress-tensor that describes matter. If gravity were to be treated in the same way as quantized fields, then the geometric tensor (on the left, in Einstein's equation) would represent the boson pole, that is, the communication ingredient, which is coherent, given the fact that the gravitational scene is also a scene observed by spectators and in which actors play. Philosophers oriented to the questions of space-time do not agree on the physical meaning of geometrical tensors. In Einstein's conception, circular causality makes matter determine space at the time that space determines matter.

$$\text{Material tensor } T_{\mu\nu} \leftrightarrow \text{Geometric tensors } \{R_{\mu\nu}, R.g_{\mu\nu}\}$$

The three Lie groups are boson generators, which might share some similarities with Chomsky's generative grammar and his universal generative structures. I suggest considering gauges as indicators of a material order associated with the quantum cosmos that has emerged. The idea of an emerging cosmos is not new and was conceptualized by Prigogine and his colleagues in the context of the physics of non-equilibrium systems. The cosmos might emerge as a transition phase. In the context of our discussion, a question arises. Are bosons generated by the physical structures related to gauges or, contrary to this, are gauges the result of boson condensation (or of the condensation of deployed fields) in the form of fermions?

I am not sure whether that question has an answer. But this does not exclude following certain lines of thought, particularly that of Wen in what concerns quantum information as a central principle for unifying matter. Fermions are not so much particles as they are embodied information. Therefore, bosons are circulating information under a different modality. The field of fermions does not necessarily possess the physical meaning that we attribute to a field. On the contrary, the field of bosons, described by tensors, becomes meaningful as a fundamental field of communications; in the same way that our natural environment does, while meeting the definition of a communication field. Following his *string-net* theory, Wen states that from the representation including qbits and entanglement, it is possible to bring out Maxwell's equation and Dirac's equation. In other words, we are referring to the photon field and the fermion field. In this way, gauges arise from a more fundamental physical process that is related to quantum information and is capable of producing fine adjustments accessible to scientists. Physical nature is not "created", but is "emergent". And if there is emergence, it is due to a myriad of basic communication and organization processes regarding stored and entangled quantum information. It is as if matter had learned to communicate.

7.4. Brief overview of matter and emergences in contemporary physics

Contemporary physics makes it possible to study and to understand the fundamental principles of emergence when it occurs or when it has already occurred in systems that have emerged and which are equipped with phenomenal functions and characters that are accessible to observation. Conversely, the theorization about emergences reveals the fundamental principles related to matter and, more precisely, to this strange quantum matter. Reality comprises two domains: the one of arranged and arrangeable structures and the one of fields that enable structures to communicate. Communication fields enable arranged structures to exchange signals which, if necessary, influence arrangement in such a way that the system is restructured by playing on matter dialectics. *In an emerging system there are deep structures that generate communicating signals. In a reciprocal way, these same signals are interpreted by deep structures which can be conceived as immergences.* The external order in the signals and the exchange of information are in accordance with information organized in deep structures (with processes that are similar or identical to those described by quantum entanglement).

The material world is both an emerging cosmos and a substance that can produce emergences. The new paradigm associates communications with emergences. It intends to be universal and analyzes all the areas of being, ranging from matter to societies. Its physical foundation comprises the whole of quantum theories, what differentiates it from the previous attempt to universalize physics in the 1970s and which culminated in the Cerisy colloquium in 1981. Scientists gathered around the concept of self-organization which had been elaborated from the thermodynamics of non-equilibrium systems. In the 21st Century, the concept of communication emergence will be central. It will be developed from a much broader theoretical foundation than that of thermodynamics. Four theorization phases place communication and the informational order at the heart of this new paradigm.

7.4.1. *Dissipative structures*

The model of self-organization uses two concepts: that of temporal evolution associated with transformations and that of the final equilibrium with its emergences. This model served as a basis for building the paradigm of self-organization which covers not only statistical physics, but all the living, the sociopolitical and the general theory of complex systems.

Classical statistical physics began with thermomechanics, with entropy as its central variable. Its second principle was that entropy increases in a closed system, which disoriented the sciences for more than a century because natural systems are open. This hampered the development of a philosophy of nature, made it problematic and even paradoxical (time arrow), because its founding assumption was a configuration that does not exist in nature (with the exception of laboratories and the adiabatic processes). In fact, the general rule is that the non-equilibrium system is supposed to exchange energy, and therefore information (or noise). Benard's columns are macroscopic structures obtained when a container filled with a liquid is heated from below. Cyclical chemical reactions and chemical spirals represent emerging structures under non-equilibrium conditions. Matter is never in a state of "pure disorder". It is always organized in compliance with the principles of the emerging system.

The theory of non-equilibrium systems does not require the use of quantum physics. Or, at least, it does not rely upon the specific quantum properties of matter. It is quite different for the study of the exotic phases of condensed matter, particularly if we take the spin into consideration. If the role of communications can be revealed by the theory of non-equilibrium systems, it is thanks to the notion of "Poincaré resonance", judiciously used by Prigogine in his last reflections.

7.4.2. *Condensed matter and exotic phases*

The physics of the 20th Century was distinctive for the discovery of phenomena generated by material components placed under special conditions established by the experimenter, particularly those of temperature or the magnetic field. These phenomena were so astonishing that they were interpreted as the "exotic" phases of matter. For a long time, the puzzling properties of superfluid helium-4 were observed at a temperature of 2 kelvin, close to absolute zero. Viscosity, the universal character of a conventional liquid, completely disappears in a superfluid, at the time that other properties appear. The fractionary Hall quantum effect has been observed since the 1980s on materials such as graphene, combining a low temperature with a powerful magnetic field. Another interesting model is that of spin liquids, with a theory developed by Landau and Fermi. This model describes the behavior of the majority of metals at a sufficiently low temperature. It also helps to describe the properties of superfluid helium-3, which behaves like a fermion "liquid" (while helium-4 is a boson "liquid").

All these phenomena occur following the principle of a phase transition. The system's features change. Some properties disappear, while others appear. These transitions make it possible to understand emergences in the context of the new paradigm which associates the fields of communication with quantum information. If dissipative systems describe the transition from an apparent disorder to a manifest order, the phase transitions of the "exotic" condensed matter move the system from ordered state A to ordered state B. In order to elaborate Fermi's theory of liquids, Landau combined two notions. The first one, adiabaticity, was derived from thermodynamics and featured a process in which there is no heat exchange; the other notion – unknown to classical physics, but very common in quantum physics – was the exclusion principle that determines the strange behavior of fermions. Quantum spin is also a key concept for describing these phenomena produced by a charged matter that communicates. Later, it will be necessary to establish how the theory of fermion liquids can be interpreted under the light of communication.

7.4.3. *Quantified fields or the emerging "material cosmos"*

Basic quantum theory describes the observation of one or a few particles of a system prepared at the laboratory, in a non-relativistic frame. Given the three quantum field theories, the description becomes relativistic and must deal with a colossal number of particles. Quantum fields describe a communication "place", featuring a propagation of signals made singular by the bosons and a gauge adjustment. The physical instances that communicate are the fermions. Quantified fields describe a kind of idealized material cosmos.

There remains a great enigma, namely the renormalization procedure used in calculations so as to avoid the appearance of infinite quantities. The procedure uses mathematical regulators whose meaning is not considered as physical by scientists. But this did not deter Feynman from detecting trickery in the success of these calculations. Then, what could be the magic trick performed by nature, unbeknownst to physicists? For the moment, we will content ourselves with the idea of an emerging "material cosmos", together with communication fields, signals, emissions, receptions and adjustments. Certainly, these fields are more theoretical than real, but they do reflect reality and make it possible to conceive matter as emergence (of embodied information), with an adjustment of communications.

7.4.4. *Exotic phases and the model with topology and entanglement*

The tensor network method has recently made considerable progress in describing these emerging phase transitions. Quantum entanglement is acknowledged. To a certain degree, it makes it possible to describe the informational order that determines emergent phenomena with their regularities. This approach enables us to define two domains: that of the fundamental quantum order and that of communication phenomena. That is, on the one hand, it accounts for deeply submerged structures and, on the other hand, it defines surface structures deployed with their phenomena, regularities and the communication between the emerging system's components.

As Orus made it clear (see Chapter 6), tensor networks are elaborated using a tiny portion of Hilbert's space, taking entanglement into consideration. This procedure should be compared with renormalization procedures, whose principle is to restrict the number of quantum transitions represented by Feynman diagrams, through the use of regulators. The most probable hypothesis is that the world of primordial information described by quantum vectors has suppressive processes at its disposal, which makes it possible to order the expression of information. This is because every expression is caused by the result of expressive tendencies constrained by the "suppressive mechanisms" which model the emerging phenomenal order. These mechanisms are known in the living and the expression of the genome and the epigenome. We can transpose this principle to the scale of quantum matter.

7.4.5. *Overview, from Plotinus' two categories of matter to quantum materiality*

The philosophy of Plotinus describes two categories of matter, one described as intelligible, and the other as sensible. This observation invites us to reflect on the resurgence of ancient philosophies in the context of contemporary physics. This possibility is not new. Schrödinger oriented his reflection towards Hindu metaphysics. Fritjof Capra imagined certain connivance between particle physics and the

Tao. However, Eastern thought can sometimes be misleading. The thesis about the two types of matter is not much more clarifying, but the diagram that emerges leaves room for duality between the matter that communicates and the matter that is arranged and ordered, with two settings: one for communications and the other for arrangements. Or better still, the submerged order of embodied information and the tuning of communications that manifest along the extent. And in the same way that embodiment works with the verb, action is associated with arrangement, at the time that functions emerge as coordinated actions, particularly as self-motivated movements in the living world.

Thus, quantum matter is composed of surface structures, communicating interfaces, and deep, ordered structures, and also a split that manifests as a duality of form and content. It is as if the material had quantum DNA, most often entangled, whose expression differs according to whether we study the quantified cosmos, condensed matter, thermodynamic transformations, elementary particles or even plain observable quantum, as the impulse or the orientation of a single spin.

A paradigm shift in the understanding of emergences is confirmed. The era of self-organization is over. We are entering the era of quantum communications and embodied, entangled, arranged information, which are source of adjustments in material emergences. This paradigm will be summoned in other analyses devoted to the living, as well as to societies and civilizations having emerged on planet Earth.

7.5. Philosophy and the physics of communicational emergence

The fundamental principles of emergence are revealed. The universe is made up of information. Some pieces are integrated and condensed in arranged structures, while others circulate through communication fields. Now, the scientific philosophy of communicative emergence deserves two additional statements:

First postulate: *a system made up of components generates horizontal emerging properties when these components are capable of communicating information, of receiving it via the appropriate interfaces, and of generating an "operational reactive network" that regulates each portion of communication and every action or arrangement. When interpretation takes place at the "semantic" level, the emergence is vertical, as it happens in emerging life or societies. There is operational hierarchy.*

Second postulate: *for a system to last, it must keep its communicational structures. In other words, an emerging system cannot be stable and sustainable unless its components continue to use efficient and coherent signals in its effects, or even a language that they can interpret.*

All things considered, these statements apply to the systems we know precisely because we are immersed in them: be it the conscience, contemporary societies or the historical worlds, which we come to know through the mediation of writings. They also apply to the cells of a living organism with molecular mediators. Thus, a new hypothesis about the origins of life follows. What is more, these statements apply to material systems, to emerging dissipative phenomena and to all the phases of matter condensed with phase transitions. As we will see, thermal noise is most often a source of disruption in communication and entails the disappearance of already emerged material structures (crystal dissolves, metal melts, etc.). It is even conceivable that the second statement, which concerns information stability, may be useful as an interpretation for understanding the effectiveness of the Lagrangian formulation. In this case, the physical system maintains its regularities as emerged and measurable because it keeps the order of information.

7.6. Emergences and molecular communication in the living

Life remains an enigma despite the colossal findings of molecular and cellular sciences. Two quasi-hegemonic models currently govern the understanding of life: genetics and neo-Darwinism. Besides, the theory of evolution has “absorbed” genetics in order to integrate it into a so-called synthetic theory. Since the 1980s, some dissident scientists and philosophers have forcefully asserted that the theory of evolution is in crisis. The most recent author is Thomas Nagel, who wrote a controversial book on the subject, with an explicit critique of modern science and a harsh verdict about the neo-Darwinian conception, which he considers false since, according to the author, it is based on a rather rudimentary materialism. While we may well believe in this critique, the author did not take into account the results about quantum matter [NAG 12]. Truth be told, the majority of the conceptions about the living do not integrate quantum physics and when this is the case, this physics remains rather incomplete in view of the stake that the enigma of the living represents. Communicational emergence should nonetheless make it possible to overcome genetics and Darwinism.

I will content myself with setting out some very brief considerations on the determining place of molecular communications within cells and between cells. The idea is to show that life has chosen communication modes (signals and interfaces) having almost universal efficiency. The stability of communication processes has undoubtedly enabled life to continue over a period of two to three billion years, and a few hundred million for animal and plant species. Life does not evolve in a shapeless and chaotic environment, but in a world that it perceives as “familiar”, at its own level. In other words, the signals that propagate through the medium are potentially compatible with their use by living systems. Cognitive life produces its environment by selecting what is necessary for it to preserve itself and to continue the scenario of the perpetuation of the species. Human life goes far beyond this; it creates a world. Heidegger clearly explained this cosmological conception of human existence which configures a world. Animals cannot configure a world because they are deprived of language and are imprisoned in their environment. The deprivation of language and the impossibility of opening up a world are one and the same thing [BAL 10]. And I would add that language and the emergence of a world are basically the same thing.

1) *The genetic code*. I will not be original in referring to the universality of the genetic code that enables communication between the deep structures of the genome and the structures superficially

deployed by proteins. Nature has selected a type of communication that makes it possible to extract genetic information and to transform it into operational forms. DNA <-> RNA <-> Proteins. The subtle mechanisms of epigenetics enable a cell to extract information corresponding to its function in the body. Conversely, we may think that information flows in the opposite direction and that epigenetic processes enrich DNA contents.

2) *Cellular transductions*. Cells have developed systems for exchanging information with membrane interfaces on which various messengers, hormones and neuromediators are fixed. On the inner side of the membrane, proteins are arranged for transmitting the signal by means of ubiquitous messengers which are present in all species. These are, for example, cGMP or cAMP (cyclic monophosphate adenosine), as well as other molecules in a small number, like the ones in the series including inositol, PIP, PIP2, etc. These molecules transmit signals with the intervention of macromolecules of a more important size than protein kinases. These transduction systems are universal. We can find them in single-cell eukaryotes such as the famous yeast used in bakery, as well as in amebae. Ions have also been chosen to serve as information transducers. When a chemical mediator is attached to a membrane receptive protein, a channel opens, enabling the ionized atoms of potassium, sodium, calcium, magnesium etc. to pass through one by one. This is not unlike a website whose server will send a suite of html codes.

Biosemiotics, a specialized branch of the life sciences, studies the communication and interpretation mechanisms of living systems. These studies are decisive because they concern the vertical emergences that characterize life (see Chapter 11).

3) *Viruses and code communication*. Viruses are enigmatic particles composed of an agglomerate of genetic material and a protein envelope, capable of replication in host organisms but unable to replicate on their own without the aid of complex genetic information management systems contained in bacteria, unicellular eukaryotes or plants and animals. The status of viruses remains undefined. Biologists do not know whether they belong to the living or not. Under the framework of communitarian emergence, viruses appear at least as complex information signals capable of transmitting

genetic information from one animal to another, from one cell to another. If the living environment is a communication field, then viruses are signals that belong to this field and operate at the mesoscopic level (this notion will be discussed in Chapter 11). These signals are most often selective and spread between individuals of the same species. Sometimes viruses move from one species to another, what makes. This is the fear of epidemiologists worried about bird viruses infecting humans and generating a pandemic.

The principle that rules the existence of these viruses is clear; however, their precise role in life and evolution still remains unclear. In particular, it is necessary to study whether these are ancestral residues that have persisted from time immemorial, and therefore function as a sort of witness for an archeology of evolution, or rather whether they are fundamental elements belonging to the biological communication field used for exchanging genetic material, aiming at a global regulation of the field occupied by living species. Viral information might then be widespread further than we think. Virophages are known for infecting giant viruses at the time that they infect other host organisms, which results in immune protection against these giant viruses. This is the sign of an obvious information mechanism. If this phenomenon is defined as symbiotic by virologists, then we can speak of informational symbiosis. In metabolic symbiosis, organisms coordinate in order to share nutrients. With viruses, symbiosis is informational. The virus shares its information with a host, which can ultimately defend itself if it feels disturbed or even "threatened" by other viral information.

7.7. Different considerations about language and the emergence of goal-oriented societies

Human societies have emerged thanks to the development of a highly efficient, even performative, communication system. Two principles have determined the human "social phenomenon". In the first principle, innovation and the perfecting of language as a necessary condition for the emergence of orderly and organized societies. When a society maintains itself in historical time, it keeps its language (and perfects it). It also retains (religious or political) symbols and performative aesthetic forms. The second principle states that form, structure and the use of signs and language are consistent with a type of society or civilization.

7.7.1. *Language, civilizations and human enterprises*

In his book *L'homme et la technique*, Spengler suggested an interpretation (inspired by Nietzsche and by Darwinism, among others) that illuminates the role of the spoken and the written language at the cradle of civilizations. Spengler used a specifically oriented notion, that of "the human enterprise" in order to explain that ancient societies known for their monumental works were determined by their "doing" rather than by their "knowing". Language is essential for establishing collective projects, and then ensuring their achievement. Language is also used for organizing social groups into classes and into specialties, so as to grant the achievement of previously delineated goals. The greatest architectural and urban achievements dating from one to three millennia before our era testify to the fact that a certain type of society emerged in Assyria, Egypt, or Mesoamerica, with the Mayan civilization and their characteristic step pyramids, which some have interpreted as a way of getting closer to the gods. These pre-Axial civilizations collapsed but their emergence cannot be dissociated from the use of a language for conveying instructions and purposes. Spengler developed the idea that a civilization is, above all, an organized system with emerging communication and manifestations. Unless they develop the "forces" that will enable them to last, civilizations may die. China had a different destiny from that of Assyria or pharaonic Egypt. This civilization, born two millennia before our era, has been able to renew itself and to perpetuate its existence through the centuries, and it has done this with remarkable thinkers. Lao Tzu, Confucius and Sun Tzu echo Plato, Sophocles and Thucydides.

Accomplished social organizations and civilizations are accompanied by architectural, political and cultural achievements. These are emerging historical phenomena which stem from collective human enterprises, welded by organizing principles. These achievements require a type of language which, according to

Spengler, enables the art of command, but not only this. Every language shows that technical or architectural achievements are not fortuitous, but are destined to serve as a means for achieving emergent (spiritual) ends. These goals cannot be counted by hundreds. They are present in all civilizations and societies, even if the forms are differentiated. While cults are diverse, worship serves an almost universal purpose aimed at maintaining social order, or at least it was so in antiquity. Bergson could have said that the universe is a god-producing machine. In reality, these gods are emergences that blossomed in a "linguistic" society, at a certain moment of its history. There is also a use of language for spiritual (divine) and worshipping ends, as well as for translating human presence. Sophocles and Thucydides are as important as Plato and Pythagoras. Power of men, Ideas of the cosmos.

A book would not be sufficient to explain how language and writing have made it possible to organize human groups by accompanying or triggering material emergences, as well as emerging goals. The language used by Plato and Aristotle is quite different from the one used by Hesiod and Homer. Sanskrit also evolved. The language of hieroglyphics developed in four phases before disappearing. Greek philosophers assigned a goal to the city, which was to orient itself towards harmony or good. With politics as a means for achieving ends. The 20th Century saw the establishment of a technocratic policy whose ends were material. As Habermas proved it, technocracy was inseparable from (impoverished) language, which helped it to emerge and to maintain it, or rather to fill the State!

7.7.2. *The myth of Babel, communicating and building*

In the myth of Babel, the bond between language and human enterprise is mentioned in a few words (*Genesis* 11). "Now the whole world had one language and a common speech. They said to each other, Come, let's make bricks and bake them thoroughly. Then they said, Let us build ourselves a city, with a tower that reaches to the heavens, so that we make a name for ourselves; otherwise we will be scattered over the face of the whole earth. But the Lord saw what was happening and feared that men would decide to go to excess, so he made them no longer understand each other and scattered them from there all over the earth." This myth perfectly explains how emergence occurred when men started to discuss together, understanding each other so as to act collectively and to work together towards a purpose. When men no longer understood each other, that is, when noise was introduced into language, then the emerging human system collapsed and the peoples were dispersed in order to occupy the earth, provoking other, less presumptuous emergences, in the eyes of the God from the *Old Testament*. The myth of Babel describes the liquefaction of societies as a result of language interference. The tower collapses, humanity becomes liquid and spreads all over the earth.

The myth of Babel serves as an excuse for describing another form of occupation, that of historical time. Sometimes people begin to speak a new language in tune with emerging societies. However, the process is really more complex. One language either covers another one or it is perfected and transformed. In ancient times, all languages evolved, Homer's Greek was not that of Sophocles or Plato. New notions appeared. Cities emerged with new political, social, religious or architectural traits. In the post-medieval Christian West, European languages such as English, German, Italian, French and Spanish sprang up and perfected themselves. Nearly a century ago, Germany saw the emergence of a new political regime, Nazism, with an alteration of traditional language produced by the regime's propaganda (see Klemperer's analysis in *LTI - The Language of the Third Reich*). These sociopolitical phenomena illustrate the close link between the form assumed by public communications and the nature of the human enterprise that emerges in history. Language and its signs accompany historical emergences. To practice a language is to configure a place in space and in time. If language is the place where Being dwells, as Heidegger said, and the Being is man, then language is not only a dwelling place in a geographical space, but also a moment in history. The historical man inhabits not only nature but also time. The religious man goes beyond time or transfigures it.

Language accompanies not only emergences, but also resurgences. As Eliade explained, the Western Renaissance could effectively take place because philologists, artists and philosophers began to study and reuse ancient Greek and Latin languages. However, in the 19th Century, a possible renaissance inspired by the East did not occur, despite the interest accorded by some great thinkers to Indian philosophies. Eliade explained that this event did not take place because the Europeans did not make a sufficient effort to practice Sanskrit. This brief note has enabled us to evoke the phenomenon of resurgence, whose study imposes the same conceptual tools as emergence.

7.7.3. *Language structures*

The linguist Noam Chomsky developed a (now outdated) theory of language in which surface structures are differentiated from deep structures. Surface structures are related to a language's performance, that is, to the capacity of signifiers to be interpreted by listeners. According to the theory of generative grammar, interpretation is due to complex transformations or operations governed by deep structures that appear as invariant nuclei and show similarities with the physical gauges whose generating nucleus produces bosons that cross the fields of material communication. What interests us is the relationship between language (as a complex structure of communication) and the emerging intention of the society that modeled it and which, in return, was shaped by these linguistic forms. The most universal goal is the human enterprise. If all languages have similarities, it is because the societies that elaborated and perfected them had one and the same intention, which was to behave as a firm with different social divisions, professions, political and legal organizations; a hierarchy made up of authorities. In other words, to develop social functions in the same way that there are physiological functions, with emerging properties that are specifically oriented to the human being who (according to Eliade's formula) is a creator of culture, as well as a religious man.

The strong link between the “universal structures” of language and social intentions brought together the great human groups that emerged after the Neolithic era. This hypothesis is reinforced by the few updated counterexamples that appear when we study the languages spoken in some remote tribes, especially among the Pirahãs in the Amazon [EVE 10]. The language practiced by this tribe was analyzed by the missionary Daniel Everett whose many published articles encountered resistance among Chomsky’s followers. However, the simplified structure of this language comfortably matches the emerging intention of these tribes which, on the one hand, constitute small social groups, a few hundred inhabitants at most, and which, on the other hand, do not pursue the ends present in historical civilizations concentrated on urban areas and whose number of individuals (tens or hundreds of thousands) enables the emergence of an “entrepreneurial purpose”. Societies of this type have languages adapted to these technical purposes, with a recursive structure (the possibility of embedding one sentence in another) that we cannot find in the Pirahã language; this language does not have a grammatical number nor numbers for counting, nor does it have quantifiers such as “all”, “each”, “some” and “a lot”. Besides, Pirahã does not have a past form, what is coherent with the a-historical existential sense present in these tribes, whereas in historical civilizations, the meaning of the past is a determining component that appears in all the founding written or retold narratives of antiquity; the stories of Hesiod, the *Genesis* and the *Exodus*, the *Mahabharata* and the *Bhagavad Gita*, the legend of Romulus and Remus, without forgetting distant China, which also has its legends and myths. Historical civilizations were preoccupied with the origins, unlike many remote tribes which, like that of the Pirahãs, are mainly focused on the present and the way of solving arising problems. Historical civilizations were intertwined with a religious conscience. To the nostalgia about the origins which was present everywhere in antiquity, we should add the soteriological dimension of fear and hope for the times to come, especially found in medieval Christian Europe.

Recursive language is adapted to the art of command. By the same token, it is adapted to the human enterprise, as understood by Spengler. Conversely, Pirahã does not have such recursivity, which becomes obsolete in a tribe that is not a firm after all, but a community of individuals evolving in a biosphere. We will find no commandments in this society. At most, we will find simple injunctions conveyed by a rudimentary language, but adapted to the purpose of the tribe who does not have a historical memory, but simply a living memory.

In complex societies, some men specialize to the point of becoming messengers rather than actors or builders. They are agents of communication in the great groups which once were the ancient cities and now are the modern states. A hierarchy is set up between a center that informs, decides, thinks and is of the "executors", whereas other individuals specialize in the transmission of information. In past times, scribes were such communicators. The Belle Époque newspaper sellers transmitted the news which had been written by the journalists to the streets of the big cities. In the 18th Century, diligences used to carry diplomatic missives. Nowadays, these specializations of the communicating society have only deepened; we live in a time dominated by those in charge of communication.

7.7.4. *Language levels*

Communication is established through signals or signs. Thanks to informational Darwinism, interfaces are calibrated so as to express and receive signals. Communication structures are universal. There is a certain hierarchy in the signs, at least as far as human language is concerned. Eliade pointed out the need to develop a dual approach in the field of the sacred and religious expressions. Understanding these modes of communication (or communion) imposes not only a philological analysis, but also a hermeneutic approach that cannot be reduced to the former and requires the adoption of a specific method.

Human language is expressed on several levels that can be embedded: semantics, semiotics and hermeneutics. Simply put, semantics determines the ordinary meaning of a set of expressions which concern not only simple things but also other phenomena, events, everyday experiences inscribed in a place or a delimited moment. Semiotics expresses meaning by putting events in common or by bringing them together under generic or generalizing terms. It connects various situations in order to provide them with meaning, or it makes actions meaningful by relating them to the past, or to goals and intentions. Finally, the symbolic expresses a reality, a truth, a value or, generally speaking, the content of a thought, by short-cutting the limits of rationality in a condensed way.

In order to complete this picture by using three notions symmetrical to the preceding ones, the Literal level refers to the chronological order, registering the events in the moment; the Allegoric level retells a story embedded in time in a derived and synthetic way, sometimes with an intention or a goal; finally, the Symbolic attempts to insert signs into a narrative, which are supposed to connect separate realities in space and to describe meaning, that can be inscribed in eternity, with transcendental, or shall we say, universal value.

7.8. Brief notes about the semantic field

The considerations about human enterprises and language explain the genealogy of vertical emergence and whose fundamental contribution is the appearance of a hierarchical system of command, of interpretation and a system of adjusted communication. As a consequence within the system, this entails the existence of singularized components capable of ensuring the centralized organization of the totality. In fact, vertical emergence assumes the existence of a structuring and central pole associated with a peripheral pole.

Submerged nuclei translate the presence of a "metaphysical formal content" into nature. Such content is not only present in the quantum world (with the symmetry of Lie-generating groups and the bosons, which make it possible for matter to communicate), but also in human societies and in their languages, which are clearly endowed with precise rules. The (infra-scopic) world of particles thus presents itself as a sub-semantic field containing generating nuclei and selected particles for embodied blocks of information to communicate and to become informed. The proton, the neutron, the electron and the photon are the objectified elements that constitute ordered matter as a device and as a pre-semantic field, to which we have to add the molecules.

For organized matter (viruses, cells and animals), the molecular pre-semantic field gradually becomes a semantic field. The components send signals that convey various informational contents, particularly intentions and orders. However, they also send indicators, traces, signs, symbols, as is the case in animal societies, then human ones, and finally the great and complex societies which are civilizations. Man lives in nature and also in a second world, that of the semantic field which he modeled with languages and works.

Every member of a semantic community sends signals and knows how to capture the signals that make sense, which show him how to act by going in the direction specified by this signal or by opposing this direction. Hegelian dialectic determines the history of men, the development of consciousness, but may also be involved in the evolution of life. These considerations invite the development of research programs whose main theme will be *the semantic revolution of emergence*, which has just been explored in these pages.

From a philosophical point of view, in reference to Heidegger, the semantic field designates the submerged nucleus which "governs" emergence, as being refers to the Being. I conclude this idea by recalling that the semantic revolution of emergence is inseparable from the revolution of time. As it was explained at the end of section 1.2: "*the field of communications is of the form/space kind, while the operational, cognitive and efficient functioning of the system's components is of the content/time kind.*"

8

The Computer, from Physics to Biology

8.1. Computer and information in the 21st Century science

Quantum mechanics is a physics lacking an interpretation for the science of basic and elementary information. Conversely, the information sciences derived from thermodynamics, Shannon's theory, the Turing machines and, more recently, the quantum computers, strive to find physical meaning, which can be easily understood, considering that their object of study is not far from abstract mathematics. *Quantum physics is looking for an information science, whereas computing theory is looking for a physics to make it compatible with nature.*

This formula summarizes two scientific issues, the first one being, on the one hand, the understanding of communications in matter. This was the topic of the first chapters, devoted to the interpretation of quantum physics in terms of communication and information. On the other hand, the second issue concerns classical information theories, computing machines, quantum computers, complex networks and the networks studied by a contemporary biology, which is placed under the epistemological conjecture of "*information looking for a science of physics*". Cells are conceived from molecules which, in fact, are chemical forms, but an emergence structure to explain the mesoscopic functions of these cells is still missing, and then we have the macroscopic functions of living organisms.

Turing's machine and the quantum computer are epistemological markers of our time, which for some 70 years witnessed the emergence of a physics of information that completes the mechanical physics inherited from the 19th Century. It was around 1750 that the great architect (or watchmaker) appeared as the symbol of the mechanistic philosophy resulting from Newton's work, whose central notion was that of force. The great architect is a technician, not to mention a mechanic. In the course of the 20th Century, genetics, as well as certain schools in cognitive sciences, placed the notion of program or computation at the center of knowledge. In the 21st Century, the great architect was replaced by the universal computer. This instrument acted upon information, but this action was no longer a force, in the Newtonian sense. While the notion of computer is illuminating thanks to its heuristic use, it can also be misleading if we take it for a true to life description of reality. In the first place, the computer is a human invention, and therefore an artificial system. Digital life uses three devices: (1) information storage on various media; (2) communication and cabled or wireless transmission of information through the Internet, USB keys, CDs and DVDs; and (3) finally, a device we no longer have to introduce, since it has become so well-known, the processor.

While the mechanistic century of the Enlightenment was puzzled by automatons, the digital age is passionate about the Turing machine, artificial intelligence and transhumanism. Calculating machines have replaced automata. It is in this new setting that the physicists who are passionate about computing are examining the hypothesis of a nature regulated or organized by quantum computers (David Deutsch and Seth Lloyd are only two of the most prominent ones). This scientific adventure started with electronic calculators, which were based on the use of a binary system. The first computers were commercialized in the mid-1950s. It was the time of Turing and of Shannon with his information theory, which aroused a lot of interest in the context of biology, especially after the discovery of DNA and its four bases of genetic coding (G, A, C and T). This epistemological framework nourished innumerable reflections about the functioning of nature. Shannon's theory was applied to biological systems by Henri Atlan.

The development of the embryo was conceived, together with the concept of "program" encoded in the genes.

The 1950–1990 period marked by Shannon and Turing is now over. A new "epistemological deal" has placed the quantum computer at the center. This unprecedented concept can make reference to a calculating tool as well as to conjectures about nature and the cosmos. Turing machines and statistical calculations are technical tools, but were not made for understanding the universe. Ellul had identified technique as the major stake of the 20th Century. Information and its organization will be the challenge of the 21st Century [DUG 17]. The failure of "Turing's machination" allows us to question the sense of the quantum computer and its validity for exploring the enigmas of nature and the cosmos.

8.2. The research by David Deutsch and Seth Lloyd concerning the quantum calculator

The arrival of the quantum calculator will not necessarily open a new technological era because of the difficulties for manufacturing this type of machine and for using it. The principle of the quantum calculator offers heuristic perspectives for questioning "physical nature" and how information is ordered. The quantum calculator does not use the classical bit but the qbit. This notion is based on the principles of quantum physics, with the possibility of having systems composed of overlapping states and effects related to the phase, in particular. A system may occupy an infinite number of states between the $<0\,|$ and $<1\,|$. The quantum calculator associates the *computer* with information which is no longer numerical (or classical, in fact) but quantum. Therefore, does the perspective it offers about nature differ greatly from the one offered by the Turing machine?

David Deutsch, a specialist in quantum *computers*, has led his research towards a "philosophy of the physical things", as evidenced by his statement: "I think it is important to envision science not as a device for making predictions, but as an undertaking that seeks to know what things are, how they behave and why". David Deutsch and Chiara Marletto developed a theory of the "physical things", now known as the "constructor theory". The authors have observed that the information used in computers is not the same that circulates in the natural state:

> "There's a notorious problem with defining information within physics, namely that on the one hand information is purely abstract, and the original theory of computation as developed by Alan Turing and others regarded computers and the information they manipulate purely abstractly as mathematical objects. Many mathematicians to this day don't realize that information is physical and that there is no such thing as an abstract computer. Only a physical object can compute things". (Conversation with David Deutsch, Interview at *Edge*, 2012)

The modern notion of physical object leads us to the natural thing and back to the etymology of the notion of *phusis,* which meant "nature" among the ancient Greeks. It is this notion of *phusis* that Heidegger questioned in his 1935 course on metaphysics, where he discussed the fundamental question of why "there are" things, rather than nothing. *Phusis* designates that which flourishes and persists, and which in the first place offered itself as a poetic fundamental experience. This small detail makes absolute sense. Heidegger considered it worth pointing out that the nature of the moderns, with atoms, material movements and other electrons, no longer had the same meaning as in the ancient times. This detail is particularly interesting for us because while Deutsch's approach does not directly lead back to ancient *phusis*, it distances itself from modern physics based on global mechanisms and information. According to Deutsch, a new physics of information must be developed, and this by separating the physical transformations into two categories: those which are impossible from those which are possible [DEU 14]. Contrary to the ancient philosophers, who described the essence of emerging things, a new question is delineated: what rules determine that a system will be able to emerge or not? Or in Leibnizian terms, which are the rules of monad compossibility?

Deutsch thinks that physical information is foreign to the formulas used in the descriptions of contemporary physics. Constructor theory meets the requirements for formulating a new physics based on the quantum mechanics revolution. It is therefore a similar approach to the one initiated by Prigogine in the 1990s. These two physicists shared the aim of shaping a new science of physics. However, for Prigogine, it is a question of describing the emergence of the time arrow by playing with quantum physics, Poincaré's resonances and adequate mathematical tools, such as those of Liouville. Deutsch is aiming at another goal, whose principle is clearly stated. It is a question of elaborating a physical meta-theory from which the laws of the physical systems can be derived by distinguishing the allowed transformations from those which are prohibited (impossible). Prigogine's formulation describes a system that evolves and is transformed while being accompanied by an irreversible time arrow. Deutsch's Constructor Theory formalizes what can be transformed and what cannot be transformed. In other words, we are referring to possible or impossible *tasks*. We are referring to "*task*" in the sense of "procedure". The idea of proceeding is also consistent with Whitehead's notion of trial, revisited in my thesis [DUG 96].

Deutsch and Marletto also consider procedure combinations and even procedure networks. This notion is important in systemic thinking. Constructor theory is based on two device categories. One of the systems produces, causes the transformation, and positions itself as the "builder". The constructor system remains unchanged, in other words, it maintains its ability to produce a transformation. The other system is transformed; it is defined as a substrate or even as a combination of subsystems and substrates. The philosophy of the "constructor" uses the notion of attribute. At the end of the construction procedure, the substrates acquire a set of attributes. A procedure is then understood as a "spectrum" of attributes (which can be interpreted as information), flowing from one substrate to another. While it is true that the notion of attribute resonates with Spinoza's metaphysics, this does not necessarily mean the "divine attribute" of the substance. Semantic questions are important at this stage. As a permanent element of nature, the "constructor" is what corresponds to

Spinoza's concept of divine attribute, and if "things" circulate from one substratum to another in Deutsch's theory, it means that these "things" are rather related to the modes of Spinozian substance.

The constructor operates asymmetrically. It resembles the catalyst agent in chemistry, in the sense that it exerts an effect without being altered, but unlike the catalyst that accelerates transformation already set in motion, the constructor is precisely what triggers the transformation. Without the catalyst, the reactants end up reacting after an indefinite length of time, but were the constructor not to be there, the transformation would be impossible. If the constructor produces effects, then who created this constructor? This is the big question that arises from Deutsch's work. Constructor theory is part of the universe of post-modern conceptions seeking to transcend not only classical physics, but also quantum physics. The constructor can be compared to a stamp that prints its seal on a wax piece, and this is done an indefinite number of times, without the stamp being altered. Can this be a reminiscence of the doublet "nature natured" and "nature naturing" used by the philosophers of nature and other alchemists during the Renaissance?

Deutsch's research has the merit of pushing computer physics and, more generally, contemporary physical theories towards a more philosophical direction that intends to account for the complicated processes performed by a set of interacting components, which surpass the limitations of statistical physics. Constructor theory will become a new science if it is able to predict systemic development and emergence for a configuration of interactive elements, better than chemistry or thermodynamics have. If this is not the case, it will remain a scientific philosophy contributing to the asking of sharp questions and endowed with a strong heuristic interest.

8.3. Seth Lloyd and quantum order in the universe

Seth Lloyd suggested conceiving the universe as a quantum computer because this is the only possibility for understanding how nature is composed of chaotic processes and orderly phenomena, beyond the laws of physics we have known so far [LLO 13]. For

Lloyd, as for Deutsch, current physical laws must be surpassed in order to understand nature, and especially this order in the universe which cannot be explained by narrow classical theories, particularly “computation” theories, solely based on binary calculators that follow Turing’s machine paradigm. After retracing the history of calculation from time immemorial to the present day, Lloyd concluded that the universe is not a cellular automaton.

The author answers three fundamental questions he poses, by pointing out that quantum entanglement is foreign to classical information. Q1: does the universe enable quantum calculation? The answer is clearly yes; the laws of physics make it possible to perform this type of calculation, not only in theory but also in practice, and this, not only in technological applications that “force” secret codes and other computer encryptions. Q2: could it be possible for a quantum calculator to “simulate” the dynamics of the universe? According to Lloyd, even if quantum calculators seem almost out of reach in terms of technical realization, their ruling principle is the possibility of making calculations and it may well be possible for nature to use them. In 1982, Feynman suggested that quantum “devices” could simulate processes more efficiently than Turing’s computer. This idea of a universal quantum processor led to the results on quantum algorithms obtained by Deutsch and Lloyd. In 1996, they showed that a quantum calculator can simulate quantum phenomena produced by several interactive particles. While the simulation of such a system with a Turing machine (a conventional calculator) requires a lot of time with an exponential increase of duration, a quantum calculator can perform this simulation within a “reasonable” timeframe. Therefore, the answer to Q2 is an enthusiastic yes. Despite the fact that no quantum theory of gravity has been found (or simulated on a classical calculator), in theory, the universe could be simulated starting from a quantum calculator. Which leads us to the last question. Q3: is the universe a quantum cellular automaton? In fact, this is possible since this “quantum automaton” is effective for simulating any local quantum system, I will nonetheless be cautious about going from local to global.

In summary, the quantum properties of nature such as entanglement make it impossible to design a universe ruled by a Turing machine. In other words, if the binary calculator cannot even simulate a local quantum system, it is just as powerless to describe the universe, its genesis and its complexity. Conversely, a quantum calculator could simulate parts of the cosmos and perhaps the entire universe. This research on the quantum calculator is rich in epistemological and ontological consequences, and paves the way for two world systems, as in the days of Galileo. (1) The system of general relativity, in which we imagine forming and deforming bubbles of the extended cosmos; turning into matter and the other way around. Gravitational forces are like elastic forces. (2) The quantum calculator works differently. Instead of "working" like a "gravitational glassblower" who would inflate bubbles, it works like an impressionist painter who projects paintballs for knitting or painting a universe, by recalculating the target of paint microbeads at every moment or unraveling before knitting.

Is nature ordered according to the principles of a quantum calculator or is it necessary to conceive another natural device whose principles and rules are not contained in quantum physics? This question arises because nature does not necessarily work with processes or principles identical to those used by man for describing and simulating this nature. That is the trap set by the "Turing paradigm", and also by other options that focus on complexity, including the most innovative ones developed with the quantum calculator as a central device. In one formula: *nature invents, it is ordered, but not necessarily using calculation methods!* Nature "manipulates" natural information, it communicates, it exchanges, it memorizes, and sometimes a dynamic order emerges. Man has not finished with the "anthropocentric curse" which distances him from universal truths. After having attributed human traits to God, he endowed nature with the features of artificial machines, from the great watchmaker to the computer. This is what we might call the "techno-centric curse".

8.4. Going beyond the theory of information: the resonance coupling principle

Information in Shannon's "classical" sense can be represented as an artificial signal that circulates from sender to receiver. Information is updated in the sender, the transmitter, and then the receiver. For example, a microphone picks up a voice and generates an electrical signal that will be applied to a transmitting antenna after amplification. The electromagnetic signal is picked up by a receiving antenna, then demodulated, amplified and applied to a loudspeaker that reproduces the voice picked up in the radio studio. This communication relies on two independent systems manufactured by the engineers. The listener must tune the tuner to the chosen radio frequency, because the antenna picks up all the radiation flowing in the radio wave range. This principle of communication can be found elsewhere, in other technological systems, such as telephone central systems, Net servers and computers.

Conversely, natural systems are not made by an "external hand". Communication adjustment emerges through molecular, macromolecular and cellular couplings in living systems. The complementarity of interfaces arises from a natural order that we can conceive in the image of structure compossibility, similar to the pieces of a game of Meccano or Lego. A natural selection process takes place with the game of communicating interfaces and fixed signals. If atoms, molecules, macromolecules, cells and living systems are assembled, it is because they can exchange information and their components memorize it. Hebb's theory describes the formation of engrams from neurons, these engrams consequently being considered as the means by which the brain stores memory. Engrams are neuron networks whose connections are stabilized and if they are, it is because they are useful, that is, the neurons that compose the engrams stimulate each other, or rather exchange signals at the synaptic level. Hebb's law states that when an A neuron is sufficiently close to a B neuron so as to excite it, and this activation of B by A is repeated, then molecular changes occur and make the stimulation of B by A more efficient over time. The phenomenon discovered by Hebb particularly concerns the reinforcement of communicating efficiency. Conversely, neurons that remain inactive decline, and gradually lose their connection with other neurons. Synaptic reinforcement occurs when two neurons are activated in synchrony. Contrary to this, if asynchronous activations occur, the connection weakens or is eliminated.

Can we make this principle general and apply it to any components or systems that can be coupled as soon as resonance occurs when signals are exchanged? Now, is it resonance that generates the coupling? Communicating resonance helps two systems to exchange information and not be indifferent to each other. This would be similar to Leibniz's monads, which were endowed with appetition, desire and affinity. Contemporary quantum physics suggests that if there is affinity and coupling, it is because there are resonances, because there is communication and information transfer. This principle seems universal and concerns the physical elements described by quantum mechanics (see Chapter 5 on bonding orbitals in quantum chemistry), every living cell, the interactions between animals in a certain environment, and finally the human being, with language resonances and the use of frequencies, especially in music. Resonance coupling is a new concept that can be used in many areas.

8.5. From the biological to the physical

Since the late 19th Century, biology has been in the "trailer" of physics. It developed as a modern science which presented itself as reductionist and mechanistic. The conception of life is incomplete, including alternative versions that combine systemic, cybernetic, classical information theory with some holistic considerations. Indeed, two major currents compete with each other nowadays [MAY 01]. Biology went astray by basing the explanation of the living on a part of physics, but at the same time relying on the idea that contemporary physics offers a complete description of "matter", and that it leaves a sufficient remainder for biology.

8.5.1. *From biology to informed physics: an original pathway offered by Paul Davies*

Physicist Paul Davies is the co-author of an article on information architecture of a living cell [WAL 16]. His hypothesis is the following one: life cannot be explained through the rules of the physical world, even if the components extracted from the living behave as material elements determined by the rules of the physical and the chemical world. Then, we would have to admit that the laws of physics are not sufficient for explaining life, which was not the case 70 years ago. Davies evokes Schrödinger and his hypothesis of the aperiodic crystal capable of transmitting heredity. While it is true that Schrödinger did not know the works of Crick and Watson, as soon as DNA and its genetic code were discovered, biologists could not help but seeing it as a confirmation of the visionary hypothesis of aperiodic crystal. According to Davies, biology is no more practical now than it was 60 years ago with the discovery of DNA. Although many mechanisms have been discovered, there is still no adequate paradigm for explaining the living. Besides, we are not sure whether physics has achieved complete self-realization, and it is this eventuality that motivates one of Davies' core research points.

This emeritus physicist proposes a revolutionary approach that develops a new physics, transposing specific notions of living systems. In fact, it is an epistemological reversal. Instead of having physics explain biology, it is biology that makes it possible to illuminate the shadows of the physical. This epistemological "direction" is opposed to that heralded by the classical path: "Rather than using physics to inform biology, one could start by thinking about biology as a means to inform potentially new physics, if indeed such missing physics exists". "A key open question is whether the concept of information as applied to living matter is merely a descriptive tool or if it is actually intrinsic to life, implying the existence of deeper principles currently missing from our formulation of fundamental physics". [WAL 16]

The ontological status of information generates a schism between biology and physics. We have to question information, a concept that most often comes as a formal substitute, and which is difficult to conceive and define with rigor. Science historians have made reference to the formal turning point at the end of the 19th Century, notably with Frege's structural chemistry and logic. Now, at the beginning of the 21st Century, we are witnessing an informational turning point. The notion of information is key to biology and becomes central in physics again, after the blunders of the 20th Century. Davies has made reference to the "informational *oddities*", for example, to Shannon's theory and the interpretation of thermodynamic entropy as an information deficit. In another context, information appears in the statistical formulation of quantum processes with the density matrix and Liouville–von Neumann's equation. This matrix makes it possible to extend the initial quantum description formalized by the state vector to a set of "physical systems". Classical statistical mechanics (Gibbs) extends to the quantum domain. Von Neumann's formulation enriches quantum description without circumventing the limitations of the statistical, and therefore, classical conception of information, though. The schism between the world described by quantum mechanics and the classical world persists. Davies' approach is singular, with an epistemological direction that introduces two new aspects. First, it takes the living world into account, not Newton's material world as a "classical bloc". Second, it does not focus on the mechanical facet (as in Bohr's correspondence, and other more contemporary studies), but on the key concept of information.

8.5.2. *Biology and networks: questioning emergence with information and entropy*

Numerous models simulate the dynamics of information. Two types of random networks are discussed by Davies. One of them is Erdös–Rényi's purely random network, the other is an invariant scale network that describes the Internet and its social networks, as well as biological networks, with scale invariance. These networks can be compared to the one that describes "information mechanisms" in the dividing cycle of yeast, interpreted as a result of Darwinian natural selection. The yeast model has points in common with the invariant network, but differs in some aspects regarding the dynamics of information. Davies extends his research to other systems, from the first self-organized networks of molecules to the sociotechnical devices present in contemporary societies. Then, there might be structures preceding the constitution of molecular devices which ensure memory, such as DNA. Living memories might emerge thanks to a substantial physical device that encompasses, frames and determines living processes.

Davies has pointed out some specific features that characterize the flow of information, network topology, and the architecture of information in complex systems. It is possible to differentiate physical networks from specific networks of living things which simultaneously emerge as a topology and as a dynamic. More precisely, information is characterized as architectural and dynamic, and therefore efficient, as translated by the notion of active information (AI). This adopts a quantized form when it is akin to entropy, in which case an information transfer is conceived as a locally specified entropy transfer (ET) between two nodes in the network.

In addition, integrated information theory (IIT) as developed by Giulio Tononi attempts to explain consciousness focusing on the brain. This is quite interesting in terms of modeling, at the same time that it has a heuristic value that has not escaped Davies. The brain is the privileged instrument for studying the emergence of topological networks and can handle very complex informational dynamics, with local as well as global transfer, architecture and order. Two categories of information appear in the IIT. One is effective, while the other is integrated. Effective information (EI) represents the entropy resulting from the mechanisms implemented in the network, and this entropy measures the distribution of resulting information. The second one, designated as Φ, represents (integrated) information that "qualifies" to what extent the whole is more than the sum of the parts. The variable Φ is useful for comparing network information at the t moment, in a state produced by the interaction of the parts and the sum of information carried by each of the parts. As unexpected as it may seem, calculations using integrated information theory (IIT and Φ) do not differentiate biological networks from physical networks. This means that according to the criteria of this theory, the living cannot be differentiated from the non-living, whereas according to the first method, there might be "informational signatures" capable of grasping the specificity of the living. These few excerpts summarize the research of Davis and his team [WAL 16]:

> "In our view, the distinctive signature of biological information management lies not with the underlying causal structure (the network topology) but with the informational architecture, and in particular how that informational architecture "interacts" with the causal structure. [...] The ensemble of SF networks in our study share common topological features with the fission yeast cell cycle network, including degree distribution, yet they statistically differ in the scaling of information transfer between nodes. In other words, although the biological fission yeast cell cycle network and ensemble of SF null model networks share commonalities in causal structure, the pattern of information flows for the biological network is quite distinct. [...] If the patterns of information processing observed in the biological network are indeed a joint product of information architecture and causal structure, as we conjecture, they may be regarded as an emergent property of topology and dynamics. In this respect, biological information organization differs from other classes of collective behavior commonly described in contemporary physics. [...] The question arises of what features of cell cycle organization lead to this emergent property. We believe that the answer lies with the control kernel, which has the requisite property of connecting informational architecture to the causal mechanisms of the network".

8.5.3. *Information and networks, theoretical limitations and metaphysical options*

These works on information pave the way for *nomological* questions with theoretical descriptions of information, and *ontological or metaphysical* questions, with the reflection about the foundations ruling exchange and the ordering of information at the most elementary level. The integrative theory of information [TON 15] aims to describe the phenomenon of consciousness using information as a central notion. Information is conceived as a set of forms shaped through perception, rather than a quantity calculated from probability. These forms tend to be composed and ordered. The central axis of this theory is the Φ function, which measures the amount of global information produced by the interaction of different parts of a system. Every ensemble which has a non-zero Φ is assigned "an amount of consciousness". The authors of this theory recognize similarities with panpsychism, which states that all natural beings possess consciousness, with varying degrees, from mineral to animal, and of course, human. Therefore, the non-zero value of Φ constitutes a double marker, indicating the presence of an emerging systemic process, while attributing a "formal element of consciousness" to the ensemble. Thus, organized matter might contain "a quantum of consciousness".

The confrontation between the approaches of Davies and Tononi makes it possible to situate the great challenges of contemporary science and, more importantly, the ontological stakes. In these two approaches, it is a question of systemic physics, systemic biology, with information at the center of the two theories. Why does Tononi's theory not make it possible to distinguish physical networks from that of yeast? Undoubtedly because integrated information theory seeks to formalize the functioning of consciousness from neural networks, with forms at the center of the approach. Davies follows a different path by making a distinction between the physical and the biological. This justifies the idea that it is necessary to invent a new physics.

Besides, Davies focuses on the architecture of information, which is what enables cell duplication, thanks to its DNA and genetic networks. It is not a question of consciousness; information is still physical, and not formal, as in the integrated theory which seeks to explain consciousness.

In summary, for Davies: *biology demands the imposition of an elaboration of a new physics to explain matter.* For Tononi, *consciousness biology requires re-examining the fundamental properties of matter.* Integrated information theory has a definite systemic interest, but presents some difficulties for understanding how consciousness emerges. It is built with the same limits as quantum Darwinism and, more precisely, follows a form of circularity in its reasoning. The effective forms introduced play a symmetrical role to the one classical islets play in the thesis of quantum Darwinism (see Chapter 4). This means that the theory of consciousness developed by Koch and Tononi deduces consciousness from elements which are already "consciousness quanta".

These few considerations have helped us to illustrate the central conjecture of emergence, with its attempts to resolve the boundaries between physics and biology. This research contrasts with the reflections previously introduced concerning the quantum computer. The starting point of this chapter is confirmed. Quantum calculation and computing seek physics to describe the transformations and the cosmos; complex biology is in search of a physics that explains the emergence of hierarchical molecular networks, some of which have cellular functionalities or functions such as reflexive thinking. Finally, we must point out that emergence cannot be dissociated from evolution.

8.6. *Kronos*, *Telos* and the evolution of living emergences

Does nature select systems capable of ordering information? It is easy to conceive that these systems have been selected naturally. In a Darwinian frame, this means that they are conferred on the species which possesses an adaptive advantage. This is what Tononi implies by invoking a natural selection that generates organisms whose integrated conceptual information, defined as Φ_{max}, is high. The reason is simple. These organizations are capable of capturing and integrating a considerable amount of information to adjust their "adaptive strategy".

Nevertheless, the selectionist explanation has its limits. The development of increasingly "efficient" systems in the organization of information is not necessarily the result of natural selection. Evolution must be explained by life and not the other way around [DUG 14]. In this sense, the progression in the order of information would rest on a property of living matter and, ultimately, on a physics of the informational order, even on information metaphysics. Life creates and organizes information. Selection might then be a selection of vital functions, not of species. Or even a selection of individuals, able to use information and to produce it. Is there an evolution determined by the physical principles inherent to living matter (which are found in the study of non-living matter), instead of being determined by the principles contained in the environment? If that were the case, life would then be a submerged and dynamic order of information put to the test by the experience in the environment, this experience particularly summoning both the technical devices to act as well as the cognitive devices to capture, to order information and to guide the action.

Evolution is currently conceived with a certain materiality determined by a single direction of time, which is defined as irreversible and which has been studied by classical and post-classical physics (Prigogine). Its principle is *Kronos*. Mutations, recombinations and natural selection are governed by *Kronos*, and directed to the future. Now, the new physics that is emerging calls for an opposite direction of time, whose principle is that of *Telos*. The next two chapters will clarify *Kronos* and *Telos*.

But to conclude, an epistemological view warns us that current considerations on physics and the living have broken with mechanism, a practical solution for avoiding the dead ends of vitalism. The hypotheses about the architecture of cellular information and integrated information announce a return of vitalism from a radically metamorphosed perspective. There is not a vital force adding to matter, but an informational order that emerges from matter and its communications. Life rests on organization rules, which are specific to living matter, in the same way that there are rules specific to the hundreds of phases of condensed matter (see Chapter 6). These living rules are not pre-existing, but emergent. This is what differentiates them from the dualism of vitalism, a model in which the soul is incorporated to matter.

9

Time Philosophies: *Kronos, Telos, Kosmos*

9.1. Time is plural, as it is enigmatic

Before approaching the question of time, a warning is essential. As St. Augustine has pointed out, talking about time is not easy. The only time about which we can speak without ambiguity is the one measured by physics. It is a time that passes by. In the context of this essay, time as something that is no longer formal but dynamic which may be related to the efficiencies and goals of nature is focused. Time would then be that which produces emergences or destroys them. This time is related to information and to energy. *Now, an unprecedented question arises, which concerns the meaning and the root of time.*

Time escapes us. In the first place, time is what separates the emission of a signal from its reception by two material elements. If the signal is electromagnetic and goes through the vacuum, then the speed is constant, equal to c. This time is spatialized. A light year is a unit of length. "Real" time is not related to space but to changes in matter. These changes are communicated to us by the interfaces of the extended world. If we are aware of the time that passes by, it is because there is a remnant of time and because the present is not instantly forgotten. In addition, memory is involved. Looking at the sun's position in the evening, we are aware of the time that elapsed if we remember the position it occupied during the morning. If, in the meantime, we saw a farmer plowing his field, a flight of cranes hovering over our heads, we heard the bells ring, we worked at the workshop, then watched a game of petanque, our consciousness of time asserts itself. It is through memory that we know that things change over time because we remember situations that were different some time ago, or even a long time ago. Our perception of time is quite trivial to understand. On the contrary, the origins of time still escape philosophy and science.

Science is endless and can pursue its unlimited technical developments without needing to make a decision, because technique progresses as if guided by a cunning intelligence. On the contrary, there is a domain determined by decision. It is the domain of philosophy. Therefore, we will try to understand what we are referring to when we talk about things called “matter”, “cosmos”, “life” and “consciousness”. This understanding can be achieved only by considering a side road leading to examine the refoundation of all modern science, starting with physics, for it is this discipline that enables access to universal realities and principles.

Our previous analyses questioned physics by exploring the world scene, the recording of events on this scene, quantum resonances making it possible to transmit information and the arrangement of things. However, on this scene, changes also occur, such as the invention of complex forms and information, biological, psychological, cultural transformations in the evolution of man. And it is the time of history, or more generally, a biographical time that is also the time of life on earth. For thousands of years, man has known that things change in time or throughout time, an understanding that increases with the knowledge about the things of nature and human narratives. This detailed and precise understanding does not prevent the fact that time escapes us when we seek to find its root.

The 19th Century was decisive with the discovery of irreversibility thanks to the second principle of thermodynamics, then with Darwin’s reflections on evolution as well as on the ideology of progress. These two scientific advances generated shocks in the field of knowledge. We discovered that species were not created in one step, but were the result of biological processes spread over millions of years. Life leads to complexification and order. However, in the physical world, the arrow of time translates an irreversibility interpreted as a race towards disorder. Thermodynamics relies on irreversible time, unlike Newton's rational mechanics, in which the film can be played forwards or winded up without the rules changing. The second principle has given rise to a misunderstanding. A closed system evolves into a state which is interpreted as disorder. Hence, "living matter" becomes more complex and evolves towards a "certain order". And if this is the case, it is because living systems are open. They exchange information and it is their "nature".

We will begin with some generalities. Philosophy has access to time through consciousness and through experience. The descriptions of time were already detailed in antiquity, which also measured time by studying celestial movements. The 14th Century saw the generalization of time measurement with more precise instruments, and it was modern science which, through Galileo and Newton, set physics in the context of mathematical time. On the world stage, the natural history of species, and then the history of men, unfolded. Time is what characterizes changing things. In fact, we should talk about time in the plural. Some hints are suggested, including two basic categories: (1) Cosmological time that extends and accompanies things perceived as immutable, like the movement of the planets or atom stability; and (2) then, there is the time that interests us closely. It is the biological or, more generally, biographical time, including the life of organisms, the life of one human and the history of humankind. Biographical time is also the one that accompanies irreversible phenomena and generates a break between the past and the future.

The misunderstanding of time comes from a conception of things based on modern science. We have to surpass and open the frames. Time accompanies change, communications and evolution towards order or disorder. Three orientations must be considered. In addition, three dynamic principles explained with ancient and modern denominations (Hegelian philosophy): *Kronos* and affirmation, *Telos* and suppression, *Kosmos* and conservation (balance, compression). Besides, we will have to make a distinction between the time of the cosmos (regulated by gravity) and biographical time, related to temporal emergences and communications (quantum, magnetic and electrodynamic processes). For the time being, we will focus on a few meanings and appearances of the concept of time, as characterized by philosophy since antiquity.

9.2. The "qualities" of time: an excursion through Eastern thought

Let us recall that, in modern physics, the deterministic arrow of classical mechanics (*Kosmos*) describes the dynamics of a system which does not exchange energy, and follows the principle of least action. The arrow of action (*Kronos*) describes the disturbance of a system by its interactions with the environment, or the disturbance of an environment by a performing system that disseminates information. The arrow of cre-action (*Telos*) accompanies the evolution towards order after a system has been disturbed by becoming informed. *Telos* determines the teleological orientation of the components.

The dual metaphysics of *Samkhya* conceives creation as material and intellectual. It involves three independent but complementary energy modes which can be interpreted as a "dynamic relationship" with the surrounding environment. These energies are designated as *gunas*, a term that specifies the qualities of matter understood as energy in oriental metaphysics. *Tamas*, *rajas* and *sattva* are the three *gunas*. *Tamas* describes the dark and inertial side of matter. *Tamas* has multiple meanings, such as slow down or impotence, the inability to start moving, and also a mode of existence without novelty, guided by the flow of forms and forces. If *tamas* characterizes an emotion, it is that of vulgar desire, of spontaneous attachment to the same thing, with a vision limited by blinkers. "Tamasic" matter is lazy, unlike the "second matter", designated by the *rajas guna*, which is passionate in the sense of it being fiery or even impulsive. In man, this quality of matter pushes him towards adventure, with the passion to go ahead, to change, which turns it into an evolutionary force. Nevertheless, *rajas* can also designate the source of violent and destructive inclinations, throwing man into uncontrolled passions and ensuing pain. *Rajas* has multiple meanings related to the consequences that it produces. It is the essence of active desire which provides the impetus for going where passion leads. *Rajas* can be interpreted as an affirmation of

Kronos, the principle of action and disruption. Let us examine the last *guna*, *sattva*, whose meaning is also multiple. *Sattva* represents the conscious, balanced, luminous and transparent quality of matter. In man, we owe it to intelligence and harmony; it is synonymous with balance, purity, even holiness. In the field of evolution, *sattva* is responsible for the cognitive qualities of living matter. Therefore, *sattva* is ruled by *Telos*, the principle that sorts, selects, orients and, above all, foresees all ends. *Sattva* means "enlightened matter", which is able to tame the ardor of *rajas*.

The metaphysics of *Samkhya* states that when one attribute dominates, it tends to persist, but sometimes the equilibrium is broken. Attributes come into motion, affecting each other to create a new state. Two principles are involved: persistence and alternation. *Rajas* is placed in an intermediate position, which can lead from *tamas* to *sattva* and the other way around. This alternation offers the possibility of cycles in the transformations of nature. *Tamas* represents the attribute of a system closed to the environment and thus related to the principle of least action in physics (and the second thermodynamic law, which leads to a particular cosmos). *Tamas* designates a stationary system, which can be interpreted as equilibrium governed by a restricted *Kosmos*, with poor information exchange. However, this order is likely to be altered by *rajas*, the principle of action creating information disturbances and noise, which is likely to disrupt a system and its environment. Finally, *sattva* is involved as a principle of order and balance, corresponding to *Telos*. This balance is interpreted as a new state regulated by a new *Kosmos*. This leads us back to *tamas* and the cycle can restart. This metaphysics of energetic qualities forming cycles has been applied by many primitive societies. It has been summoned during the rites of regeneration, and it has promoted the idea that humans can also participate in the process of creation and generate their own small-scale cosmos. These sacred considerations were cleverly summarized by Eliade.

Vedic metaphysics and Greek thought are in connivance, particularly in what concerns the cyclical conception of time, which has united ancient Greece and India. The energetic qualities appear in the allegory of horses used by Plato in the dialogue of the *Phaedrus*, in order to describe the human soul. The soul is composed of three parts. It is similar to a coachman driving two horses, one good, and the other spirited and vicious. The virtuous horse expresses sattvic attributes, while the disobedient horse, untimely and obscured in its discernment, expresses the rajasic attribute. If we make reference to Plato's concern for orderly things, then this allegory describes the moral and aesthetic qualities of the soul, depending on whether it masters time or whether it is doomed to the whims of time:

> "In the beginning of this tale I divided each soul into three parts, two of which had the form of horses, the third that of a charioteer. Let us retain this division. Now of the horses we say one is good and the other bad; but we did not define what the goodness of the one and what the badness of the other was. That we must now do. The horse that stands at the right hand is upright and has clean limbs; he carries his neck high, has an aquiline nose, is white in color, and has dark eyes; he is a friend of honor joined with temperance and modesty, and a follower of true glory; he needs no whip, but is guided only by the word of command and by reason. The other, however, is crooked, heavy, ill put together, his neck is short and thick, his nose flat, his color dark, his eyes grey and bloodshot; he is the friend of insolence and pride, is shaggy-eared and deaf, hardly obedient to whip and spurs." (Plato, *Phaedrus*, 253)

The qualities of human time are associated with desire that spins according to the flow of information received or created by the human soul, which knows how to strive to desire things in an indefinite number. Ultimately, it is the information control system, which does or does not regulate desire. Civilizations and societies have not ceased to control human desires so that collective life is not chaotic, a prisoner of individual drives and selfish passions. Eastern philosophies see the temporal world with its suffering caused by uncontrolled desires. Therefore, Eastern wisdom consists of getting rid of those desires and even the wheel of time responsible for desire, frustration and other fatal passions. In the Christian West, the doctrine of the original sin engaged morality. Wisdom was replaced by salvation, guided by the quest for morality and good. It was not so much a question of going out of time, but of transcending it, or at least warding off its negative effects. Thus, philosophical thought came upon the question of evil, for example, Bataille and his writings on literature and evil. The duality of good and evil is specific to the West. Apart from Zoroastrianism and Manichaeism (recycled in Western thought), the East does not conceive things in the same way. In the ancient Greek world, which was under the influence of the East, bad behavior was not linked to a deficit of good or an excess of evil, but to the absence of illumination and vision, in other words, a lack of truth and of reason, as indicated by the Platonic allegory of the horses. Or an excess of power in Thucydides' thought.

9.3. Aristotle

Aristotle's physics contains a conception of nature (*physis*) that we, modern, could define as qualitative. Besides, Aristotle considered that mathematics contributed nothing to the understanding of natural things (what differentiates him from Plato). Nature as understood by Aristotle contains in itself the principle of the motion of things, as well as the finality of things, which reach their natural destination and are transformed according to the rules fixed by nature, with the principle of entelechy and having the soul as formal determination, since this is the form of a body potentially harboring life. The natural philosophy of the ancients is based on an "empathic observation" of the thing, or rather "embodied" cognition, achieved by an observer intertwined with nature, this observer making "body" with the perceived thing which, once interrogated, delivers a definitive answer. This makes it possible to establish the natural tendency which, as it was clearly explained in book VI of Aristotle's *Ethics*, represents truth. Once the thing is founded in truth, there is nothing else to question. The conception of the world in medieval times accentuates the ancient diagram by transcending it thanks to a theological framework: essence, substance and accident. Divine order as opposed to the disorder of things. Things also have a place, assigned not by nature, but by God. As Koyré put it, the system is closed.

Aristotle thought that time is the measure of motion. But what is motion? "The actuality of what admits of qualitative change, qua admitting of qualitative change, is qualitative change; of what admits of increase and decrease (there is no common term to cover both), it is increase and decrease; of what admits of coming-to-be and ceasing-to-be, it is coming-to-be and ceasing-to-be; of what admits of locomotion, it is locomotion." (Physics, III, 1, section 8)

If we understand Aristotle correctly, there are three "effects" of time. That which is responsible for change and corruption, the one that produces growth and generation, and, finally, the one that is associated with motion, in other words, moving a mobile. The corruption of things refers to disorder and deductions made from the second principle of thermodynamics. There we see the dominion of *Kronos* and the erasure of the guiding principle (*Telos*). Motion refers to the classical mechanics of material arrangements (Newton and Lagrange) and to a temporality linked to mechanical determinism. Phenomena are then conceived as a film that can be played backwards. The third category of time accompanies the development which, according to Aristotle, comprises two movements: increase and decrease. The development reveals the *Telos* arrow and to speak in modern terms, it is the arrow of complexity, of order, of self-organization. Aristotle completes and refines what was already thought by Heraclitus and by "motion", we certainly have to understand "change".

In his treatise *On the soul*, Aristotle resumes the distinction outlined in *Physics*, but this time he examines the movements of the soul. Thus, a key concept emerges, that of entelechy, which designates an essential feature of the soul, which also poses some translation difficulties. Entelechy refers to form, which is consistent with the definition of the soul as the form of a body potentially harboring life. But entelechy has other meanings, as well. The entelechy enables form to achieve its own end in itself. When an entelechy describes the development of an organized body, it also acquires the meaning of perfection, of achievement. Finally, entelechy is also seen as a systemic principle, playing an organizing and coordinating role when a living creature is composed of parts. Aristotle thus describes the different parts of a plant, and then defines the soul of the plant as the entelechy of an organized natural body.

Hence, the concept of entelechy contains two determining principles that explain the living and the existence of beings from a biographical point of view. "Entelechy" then means "direction" in the sense of *telos*, "orientation", "achievement" and secondly, "organization", "coordination". Nevertheless, entelechy does not make it possible to reach the understanding of emergence, only that of morphogenesis. Starting from Aristotle or Leibniz, we cannot truly access a Darwinian conception of evolution.

For the idea to be complete, let us observe that the four movements described by Aristotle make it possible to envision a biographical time, that is, a chain of events with a beginning, a development and an end. This can refer either to the motion that carries a mobile from one place to another or to the processes of increase, decrease and corruption. These determinations are those of the sublunary world, to which Aristotle adds a superlunar world with the stars, regulated by a fixed order, that of *Kosmos*. This celestial world is described and explained in *On the Heavens,* in which we observe the use of an enigmatic concept, that of motionless motor. This means stationary movement, as this motor is linked to the fixed stars, while it transmits motion to the planets. If we can delineate an idea of time, it is irrefutably that of cosmological time. The superlunar world is regulated differently from the sublunary world. The physics of the 21st Century has unified the worlds, but it has not solved the dilemma of the two notions of time: a biographical and a cosmological one.

9.4. Leibniz

Nicolas de Cues, priest and scholar of the 15th Century, offered a concept of natural things with two poles, *explicatio* and *complicatio*, which can be translated as "development" and "envelopment". This conception undoubtedly inspired Leibniz, a Baroque philosopher with his monads made of folds, enfolds and unfolds according to the formula suggested by Deleuze. Three centuries later, the quantum physicist Bohm took up this idea by claiming a duality between implicate and explicate order. The internal and monadological order is consistent with the possibility of an emerging complexity based on the memory of the monad, which describes matter following a different conception of atomism. Thus, matter is not just a collection of objects with external properties. It is also a substance endowed with interiority, memory, received (and if necessary, ordered) information, without forgetting appetition, desire and affinity. Organized matter in the living world is guided by a *telos*, and once again, we come across this notion of entelechy, which was widely used and revisited in Leibniz's *Monadology*, which we will briefly quote:

> "All simple substances or created monads may be called *'Entelechies'*, or created Monads, because they have in themselves a certain perfection (*échousi to entelés*). There is in them a sufficiency (*autarkeia*) which makes them the source of their internal activities, and renders them, so to speak, incorporeal automatons" ([LEI 91], section 18).

> "Thus although each created Monad represents the whole universe, it represents more distinctly the body which specially pertains to it and of which it constitutes the entelechy. And as this body expresses all the universe through the interconnection of all matter in the plenum, the soul also represents the whole universe in representing this body, which belongs to it in a particular way" (*Ibid.*, section 62).

> "What we call a *living thing* is a body that has a Monad as its entelechy or its soul, together with that entelechy or soul. And we call a living thing an *animal* if its entelechy or central monad is a soul. Now, this body of a living thing or animal is always highly organized. Here is why: the universe is regulated in a perfectly orderly manner; and every monad is a mirror of the universe in its own way; so the representing monad must itself be orderly; so the body that it represents (thereby representing the universe) must be orderly" (*Ibid.*, section 63).

> "We can see from this that every living body has one dominant entelechy, which in an animal is its soul; but the parts of that living body are full of other living things, plants, animals, each of which also has its entelechy or dominant soul" (*Ibid.*, section 70).

> "The necessary elements for every created being are: a composed body, and a central monad or entelechy. [...] And it is by representing more distinctly the body that is particularly affected to it, that the entelechy or soul represents the whole universe. Thus, there is no soul without a body in nature" (É. Boutroux, *Notice sur la monadologie*, III, section 2).

These excerpts confirm the meaning of entelechy chosen by Leibniz in the sense of an organizing and substantial principle designated as Monad. An organized body is composed of peripheral monads and a central monad that makes the whole autonomous and perfectly regulated. The monad is equivalent to the concept of submerged nucleus that we introduced earlier. In the philosophy of Leibniz, there is no emergence, but a development of organisms conceived as a "*kosmos* of monads". In this way, each monad is considered as a reflection of the universal cosmos. In this diagram, the creative aspect of time and *Kronos* as something which enables the transformation and emergence of species are still missing. In Leibniz, the *Telos* of the monads works with *Kosmos* (God), but after Darwin, *Telos* is definitely associated with *Kronos* and emergences. Thus, we move from an antico-medieval conception of time to a Darwinian and Hegelian one. *Telos* is no longer positive (pre-established direction), but suppressive or negative. To a certain extent, *Telos* is a moderator of *Kronos*, playing a decisive role in the organization of emergences. The post-Hegelian era is that of the open time anticipated by Leibniz and his *Theodicea*, which was later embodied in the philosophies of progress and History, as regards Spirit, and Darwinism, as regards Nature (the thermodynamic arrow and its enigmas not to be forgotten). The contemporary world after 1815 was placed under the aegis of two "openings" of time: on the one hand, the Hegelian conception of time which involves History and the Spirit, and on the other hand, the time of evolutionary species, with Darwin.

We can note the great difference between what *Telos* can mean in the entelechy that appears in the monadology and *Telos* in the Hegelian dialectic. In the first case, *Telos* means a direction towards achievement, while in the second case, it means the suppression of the orientations which do not comply with the reason propelled by *Kronos*, or rather with the dialectical resolution of a particular situation at a given moment. This observation leads us to the last great philosophy of time, with *Dasein* and a possible resolution of time which we will define as non-dialectic in Heidegger.

9.5. Being, time, things

Heidegger used to explain the history of Western philosophy by placing in its center the oblivion of Being. What if the history of Western science was based on a form of time forgetfulness? Being was forgotten because it was robbed from thought, it remained hidden, concealed. With modern physical sciences, the notion of time becomes a subjective experience that can be measured. In fact, time has become a parameter in theoretical formulas, describing dynamics (forces) and kinematics (trajectories). No physical theory can ignore time. Except for Shannon's information theory, for example, which is not actually a physical theory. If we can criticize science for an omission of time, it is only in the sense of a misapprehension. There is a category of time that seems to have been particularly set aside by science. This category is none other than *Telos*, which ultimately led to a scientific conception of nature that banished final causes, as well as all doctrines including, albeit lightly, the notion of finalism.

After Galileo and Newton, and later with the enigma of entropy in the late 19th Century, modernity focused on two categories of time whose principles are *Kosmos* (thermodynamics) and *Kronos* (rational mechanics, cosmology). The third efficient category of *Telos* has remained unknown to science, not to say completely absent from it. *Telos* was reduced to the *Nomos* of evolution towards order. For a modern intellectual, physical nature is blind. Nevertheless, order only occurs under conditions of openness and non-equilibrium. Disorder can arise from order according to the scientists of the systemic paradigm, centered on the concept of self-organization. And time may have eluded science.

Heidegger sought a path in order to retrieve the sense of Being. But instead, it was rather the question of time that he developed. *Dasein* is somehow the scene where all beings meet. The philosophy of *Dasein* elaborated by Heidegger must be interpreted as a cosmology. The fundamental problem of Dasein's analytics is how to access Being on the basis of existence, that is, how to rely on biographical time in order to access cosmological time. Biographical time is linked to the consciousness of a being destined to death, ultimately understood as definitive suppression, not as achievement. Biographical time is also the time of History. In his essay *Being and Time*, Heidegger inquired about historical time by engaging in a dialogue with Hegel, from paragraphs 67 to 77. Hegelian philosophy can be interpreted as a way of moving from historical and biographical time to a cosmos understood as the realization of History and achievement. Heidegger perfectly understood the trap of Hegelian ontotheology, whose principle was to aim for the coincidence between biographical and cosmological times, in other words, to rejoin the doublet *Kronos-Telos* with *Kosmos*.

The other investigation with cosmological implications concerns the thing and the Kantian moment that follows and echoes the mathematization of the world initiated by Galileo and Newton. In what ways are these two enquiries related? They describe a kind of dramatization: that of History and the incarnation of things on the world stage. But how do things play on this stage? How does man manage to take the stage and change his world? The historical analysis of *Dasein* tries to elucidate how a world opens up, how it occurs, with its historical figures and the reunion of actors determined by authenticity, oriented by sense. The other question concerns nature. Do things come into play and does man try to understand the rules of this natural game? How do things behave, how can we understand them, where do they go? How can we question these things and try to grasp the place they occupy, the role they play (if they play a role at all) and thus why are things the object of knowledge, and not formless or aimless foam? Thus, we understand why the appearance of things on the world stage and their dramatization by man pave the way for a double interrogation, one concerning the stage and the other regarding time. This notion of stage is not common, neither in science nor in philosophy, or at least it has not been designated in this way. In Newtonian physics, space-time is a fixed scene where things happen. In Kant, we also find the idea of a stage, and we certainly find it in Heidegger where, thanks to the *Dasein*, the scene acquires certain thickness and a temporal opening (what reminds us of the exodus from Egypt, narrated in the *Exodus* on the *Old Testament*). In Newton, time has emerged, for Kant, time is a transcendental submerged *a priori*. For Heidegger, time is associated with the power to emerge (*Ereignis*).

Therefore, the world of scientists has two cosmologies: a scientific one, as established by Einstein, whose principle is the equivalence between mass and geometry; and the other, philosophical, sketched by Heidegger and whose purpose is to bring out what is hidden as the truth of Being. The confrontation between Einstein and Heidegger has not begun yet. It will develop around the question of time. I would like to add two remarks before we finish.

Relativistic cosmology uses a time parameter which is that of astronomy, a cosmological time that can also be described as universal. Contemporary cosmology deduces a biographical time from universal time, by elaborating the Big Bang Theory and attributing an age as well as a history to the universe. The thermodynamics of the black hole also teaches us how biographical time can emerge. The quantum black hole has a history and sometimes an end; it eventually evaporates and disappears after having radiated. Contrary to this, the analysis of *Dasein* begins with biographical time, the time of experience and History, and seeks to understand the cosmological time of Being. Finally, let us observe that *Dasein* can help us outline a philosophy of emergence, with *Dasein* as the first instance of consciousness. Only the future can tell where these considerations will lead.

9.6. Things in modern science

Aristotle used to conceive of nature as a hylomorphic substance that determines the motion of things on a stage to a place assigned to them, among other. The setting determines things. For Newton, movements and positions become properties of the thing, which can be measured and described by mathematics. With celestial mechanics, the ontophysical status of bodies is radically altered. The place is no longer the place to which the body belongs by virtue of its intimate nature, as Aristotle thought. In fact, the body occupies a position that is acquired at every moment in relation to other positions and other bodies. In rational mechanics, the Lagrangian contains all the information which determines the position and the motion of the thing represented by a material point.

Heidegger noted that mathematics belongs to the thingness of things in modern physics, and recalls the subtle distinction between mathematics and dynamics made by Kant in paragraph 11 of the Analytics of Concepts (CPR). Mathematical concepts belong to the intuition of the thing, while dynamic concepts relate to the existence of the thing. Kant stated that this distinction is not related to the thing itself, but to the understanding of it. This might suggest that modern knowledge does not concern the thing, but the way in which the scientist projects a reading grid drawn from his own background, onto the thing. Modern science measures and calculates the staging of things according to the experiences and forces that change the arrangement of the masses in relation to each other. This interactive game eventually inspired Mach's principle of the relative arrangement of things, which ultimately hinted that there is no external instance, prime engine or otherwise, that places things where we find them. The Newtonian principle of the insertion of mathematics into the thing finds its culmination in general relativity, with the equivalence between geometry and matter. This achievement certainly did not escape Heidegger:

> "As a result, the general concept of nature is modified. Nature is no longer the internal principle from which results the motion of the body; but rather, nature is the modality of the diversity of the changing relationships of the bodies' positions, the way in which they are present in space and time, which in turn, as regards possible areas of order position or determination, do not have any proper mark anywhere." [HEI 67, p. 99]

Now, let us examine the indeterminacy of order and a question regarding the principle of *Kronos*. Does quantum irreversibility bring about classical irreversibility? This question poses a major problem because time is a classical notion and as such, is no longer relevant to the quantum phenomenon. If time appears in quantum mechanics, it is because this physics has incorporated some classical elements into its formalism, since they are indispensable for inscribing quantum observation in our classical device. In the same vein, entropy (an amount used for describing irreversibility in the classical sense) should not make part of quantum mechanics. The classical world does not explain the quantum world, but conversely, if we start from quantum and we follow certain assumptions, we will reach the classical world and we will even understand it with more fundamental details. In the 1980s, Prigogine attempted to introduce a microscopic entropy operator for describing irreversibility in quantum mechanics. It was a legitimate idea, since many classical magnitudes have their correspondence in quantum mechanics (for example, the impulse that becomes an operator). However, it is not certain that the transposition of entropy to the quantum level is legitimate. Prigogine seems to have abandoned this idea in the 1990s (read *The end of certainty*). There, he took the opposite path, striving to build a classical universe starting from quantum. A universe including both the emergent cosmos as well as the "world" of classical matter, with its materials, solids, liquids and gases. And which is transformed by following the laws that lead to the formless equilibrium of closed systems and towards the emergent and stationary forms of open systems.

Quantum physics shows that "matter" is a substance that exchanges information. The monadological principle directs matter towards association or dissociation, according to the information exchanged and the resonance with the instances that interpret this information. In an open system, information is renewed and the system gradually or suddenly creates the efficient instance that orders its components. It is as if the instrumentalists created the conductor who will organize the music and the dynamics of the ensemble. A closed system evolves towards cacophony and informs it, while an open system can evolve into symphony and forms. The appearance of these organized structures places us face to face with the enigma of time and creativity in nature, a creativity that constitutes a cosmological category, paraphrasing a statement dear to philosopher Whitehead, and evoked by Prigogine. In this sense, we can speak of the principle of cre-action.

In summary, the thermodynamics of non-equilibrium is suitable for a superficial study of nature. However, it is quantum mechanics that best describes the systems open to information exchange, thanks to the quantum resonances related to the undulating essence of "material substance". With quantum physics, we gain access to a new understanding of the world scene. Matter and the cosmos cannot be reduced to bodies that move and clash. Nature is essentially a space of communication, as much as of arrangement. If we accept this distinction, then we will begin to understand the question of time. Because there are at least two times, the one that is related to communication and extension (interfaces and communications), and the one that accompanies (or even determines) the changes in the arrangement of things, that is, the transformations of matter, of the cosmos and of life related to the information "from and inside" matter.

The Kantian analysis of the scene of phenomena is thus summarized by Heidegger: "Space is the form in which all phenomena occur. But time is not limited to these, it is also the form of internal phenomena, that is, the emergence and continuation of our attitudes and our experiences. Hence, time is the form of all phenomena in general [...] It is not time that changes, but something that is in time" [HEI 67, p. 237]. We can surely note the enigmatic character of this time as an *a priori* form of subjectivity. But for Kant, time is not a driving force of nature. Rather, it resembles a reel projecting the successive images of a phenomenal film.

Kant's physics does not blend into Newtonian mechanics. More exactly, Kant questions not only things but also their reception and their perception throughout an extended scene, taking into account the results of modern physics. In Newtonian mechanics, space-time is a separate scene on which things are written. Besides, the observing subject is progressively retracted from the objective scene, which becomes no longer related to a subject but to an observer by means of instruments, which are also objects belonging to the theater of things and phenomena unwinding on the stage. Conversely, in Kant's transcendental aesthetic, things, time and space are, so to speak, reunited, or at least, intertwined. This also means that the subject is intertwined with the scene on which the phenomena unfold. Kant refuses the retraction of the subject. This description is radically alien to Newtonian physics with its masses, forces, accelerations and trajectories mapped over a mathematical (geometric) extent.

9.7. Hegel, Nietzsche, Heidegger: three prophets of time

The contemporary era saw the rise of science and the erasure of philosophy in the field of knowledge about nature. The contemporary era was also the moment in which European history was unleashed and become global. Three post-Kantian philosophers placed time at the heart of their reflections, often with History as their object of study, or if necessary, as an object for triggering thought. *Kronos*, *Telos* and *Kosmos* became universal principles both immanent and transcendent to the course of human affairs. With each philosopher, there appeared a central notion: the Spirit for Hegel, Power for Nietzsche and *Dasein* for Heidegger.

For these three philosophers, time must be considered as consubstantial with the emerging transcendental modalities which are the spirit, power and *Dasein* (as an emanation of the call from Being to consciousness). This time cannot be confused with that of physical or biological emergences, although they all share a biographical essence. A common denominator brings these three intellectuals together. It is the consciousness, a high and deep consciousness which in Hegel becomes spiritual, whereas for Nietzsche, it is Power that becomes conscious and finally, in Heidegger, consciousness is founded by *Dasein*. If we consider that consciousness is a communication (or communion) device, then the philosophies suggested by these three thinkers can teach us something about the notion of emergence and how this has manifested throughout the existence of the human subject. Among these emergences, *Telos* represents an instance that determines an orientation for an existence whose efficient cause is embodied by the principle of *Kronos*. The impact of *Kronos* is weighed differently by each of the three philosophers.

The study of these three singular philosophies, to which we can add the phenomenology of Husserl, makes it possible to articulate three fundamental concepts: communication, emergence and time. Consciousness is based on communication (or communion) with oneself and with the world.

9.7.1. *Hegel and the dialectic resolution of* Kronos *with* Telos

Hegel's philosophy is organized around central notions, especially teleology. The figures of *Telos* appear in several texts. The best known is the *Phenomenology of the Spirit*, with its introduction where absolute knowledge is announced as the fulfillment of consciousness, together with the enigmatic notion of the recollection of spirits, or rather, a recapitulation of man's spiritual way of the cross. The other text that evokes *Telos* is *Reason in History*. The idea of the cunning of reason, as well as the thesis of the end of History, is no doubt teleological. Its result must be the rational State, which represents the most perfect historical organization of mankind. In other words, a political cosmos guided by *Telos* and self-realizing in *Kosmos*. The emerging order becomes a stationary emergent order. Teleology is also mentioned in a decisive chapter of volume 2 of the *Science of Logic*. This text describes the progressive organization of a system in three stages: first, that of mechanisms and the indifference of the elements, then chemism, which can be interpreted as the moment of communication and information exchange, and finally, teleology, the moment when finalities emerge.

Heidegger clearly interpreted Hegelian philosophy and its approach from the perspective of metaphysical logic, including the untranslatable notion of *Aufhebung*, which means "affirmation with suppression and preservation", without any of the three terms having a negative connotation. Therefore, we find the affirmative *Kronos*, then *Telos* as a negative process of determination with selection of meaning and orientation, and finally preservation, which refers to a "metaphysical" equilibrium, or rather a (dialectical) resolution by *Kosmos*. The concept of preservation is certainly based on a configuration of compressed information related to a time that has elapsed, to a biographical time.

9.7.2. *Nietzsche and* Kronos *transmuted into* Telos

The determination of time designated by *Kronos* is associated with the quantum of action, which adds information, generates noise and constitutes a source of chaos, in short, it informs about change. If *Telos* intervenes when the conditions are reached, change is ordered and a formal emerging order manifests. "One must still have chaos in oneself to be able to give birth to a dancing star". This famous quote from Nietzsche illustrates the efficiency of the creative principle in man and in nature, a principle of power and therefore related to *Kronos*. But in this thought, *Kronos* is somehow tamed and even overturned in terms of orientation; it transmutes into *Telos*. Many of Nietzsche's texts are interpreted in this light, particularly the considerations developed in the genealogy of morality.

The man who has a dominant instinct also holds the privilege of responsibility and knowledge about the freedom that accompanies it. This sovereign man calls this dominating instinct "consciousness", what enables him to be warrant of his own way and to proudly say Yes to himself! Although this type of awareness is emergent, its process takes time. It is not only a mature fruit, but also a late fruit (Nietzsche, *Genealogy of Morality*, II, sections 2 and 3). Nietzsche's late philosophy has power as its central theme, as well as the description of this power becoming aware of itself and managing to find a direction (the path). In short, *Kronos* is transmuted or metamorphosed into *Telos*, and it is the Nietzschean enigma of becoming that emerges together with the attributes of Being. Power is also interpreted from the perspective of Compressed Information, with a twist of the path described by Hegel.

9.7.3. *Husserl, the world and* Kosmos*-based objects*

The phenomenology elaborated by Husserl shed light on the dynamics of the three principles of time, especially thanks to a subtle analysis of the intimate consciousness of the present time which sets phenomena up on the stage (in the way the cosmological time of relativity does). This present time of *Kosmos* is comparable to a consciousness that combines what has happened with what is expected, or rather, what is pushed by *Kronos* with what is collected by an anticipation-governing *Telos* (a notion used by Kant, as well). It was by following this analysis that Heidegger was able to sharpen his thesis about man as a configurator of the world, envisioning him as a source of the conscious cosmos and as nucleus of this same cosmos. It is thanks to the present that this world is configured, and the synthetic unity of *Kronos* and *Telos* is achieved, in a similar way to the physical cosmos configured with a specific time (which have been interpreted as unit, since it corresponds to the deterministic and reversible time of classical physics). Nevertheless, the idea of a configured world goes beyond the setting up of phenomena in the present, because language is also a useful tool for configuring an emerging and specifically human world; in fact, animals do not possess language. This configured world constitutes an authentic emergence, specific to man, in that he needs to borrow the (temporalized) phenomena of consciousness and language in order to achieve it. In other words, we are practically reaching the fundamental role of communications for emergence to take place, but this time from the point of view of the human experience.

The other interest of phenomenology lies in the characterization of the object as noesis and as thought production correlated with the intended (or aimed) object, that is, correlated with the object as communication signal for the consciousness connected to its essence, which can therefore be interpreted as the unveiling of a submerged nucleus (the equivalent of the gauge in physics). This phenomenological operation is carried out by the suspension of judgment, which amounts to getting rid of the efficient movements determined by *Kronos* in order to allow the objects to fit into a world set up as phenomenological cosmos. Husserl describes the emergence of the contents of consciousness.

9.7.4. *Heidegger and the quest for* Kosmos

The famous thesis about the forgetfulness of Being serves as an introduction to the essay *Being and Time*, in which Heidegger attempted to elaborate a path for thought which would reach its accomplishment through the sense of Being. Heidegger went on a quest for *Kosmos* based on *Dasein*, a concept foreign to pre-Socratic philosophy in which the interrogation about Being had first blossomed. Human consciousness "embodied" in Dasein is intertwined with essential Time. According to Sloterdijk's formula, if Nietzsche is the thinker on stage, then Heidegger is the thinker inside the scene. Nietzsche contemplated himself as the power of being dramatized in the world, while Heidegger placed himself as open consciousness, a scene moving forward in time and making it possible to question Being. *Dasein* is waiting for the call to get onto the stage. Consciousness tries to understand what remains and sustains the world when the actors and the scene emerge and hold the "world together"; Being is like a presence knocking on the door of *Dasein* without exercising any intention or pressure. This possibility closely resembles Aristotle's motionless engine, only with a difference in size. This engine is not in the eternal cosmos (superlunary world), but insinuates itself to *Dasein* (sublunary world).

Being appears as a retracted emergence. Now, the question is not so much about a fixed expectation as in the intimate consciousness of time, but rather expectation in the search for resolution. *Telos* then appears as an event that retracts face to a lack of resolution. This existential equation has the possibility of being solved. The notion of anticipatory resolution is important and supposes an absence of will, that is, it appears as a "suspension" of *Kronos*. *Dasein* is projected in silence. It arises in the same way that Einstein established cosmological equations for which there are several metric solutions. In Heidegger's *Dasein* cosmos, the solution is not measurable, but reveals a kairological time that signals the end of expectation. *Kairos* is a time that can be filled in two ways. We can already observe something important. It is this notion of filling up time that perfectly fits the notion of immergence we have developed in this essay. In anticipatory resolution, *Dasein* should be understood as an immergence to be solved with the fulfillment that occurs on both sides of the "existential mirror", either by myself or by the other. And in the second case, we are in the Greco–Pauline conjecture of Parousia, which does not mean what happens but what takes place as an event. What arrives takes place in the world of determined things and regards us (the train is arriving at the station, Jacques has arrived for dinner). However, what happens is not determined. It is an authentic emergence in our existence. And which should be understood in relation to the eschatological time that functions as a mirror image of immanent kairological time. Or rather, to use Françoise Dastur's interpretation, the relation to Parousia does not imply "in expectation of", but "to be presently awake".

In order to complete these reflections, a word on historicity, which emerges as a set of individual fates. This historical Destiny of men is not a simple addition of fates, in the same way that the whole is not equal to the sum of the parts in a complex system. The historical Destiny is built by the common fight for setting up a collective power, particularly thanks to communication, which makes it possible for it to be shared. In the paragraph on the fundamental constitution of historicity, *Telos* appears as irresolute about care (founded by temporality), then is resolved by Destiny with the delivered (or resolved) power of *Kronos*, which in some way fills up historical time.

These few excerpts serve as a conclusion on the theme of history as emergence (a complete study would require a detailed essay):

> "The being of *Dasein* was defined as care. Care is grounded in temporality. Within the scope of temporality we must accordingly search an occurrence that determines existence as historical [...] Destiny is not composed of individual fates [...] These fates are already guided beforehand in being-with-one-another in the same world and in the resoluteness for definite possibilities. In communication and in battle the power of destiny first comes free. The fateful destiny of *Dasein* in and with its generation constitutes the complete, authentic occurrence of *Dasein*" (Heidegger, *Being and Time*, section 74).

9.7.5. *Emerging* **Logos** *and* **Telos** *in the end times*

Are the three determinations of time and emergence destined to be reunited? *Kronos*, the principle of power, *Telos*, determination of orientation and *Kosmos*, accomplished and permanent order. Power creates and is ordered once time has elapsed. *Telos* appears as a median term linking *Kronos* and *Kosmos*, in other words, as the power that finds its way towards order and perpetuation. What could be the common ground for interlacing these three poles of time? Does the philosophy of time lead to Mystery? *Kosmos* is co-generated by *Kronos* and *Telos*, the great Mystery of time, which we will not confuse with the speculative mystery of gravitational time in physics and cosmology. The *Kosmos* refers to the natural and almost immutable order praised by Pythagoras or Lao Tzu in his *Tao Te king*. Or Anaxagoras's *Nous*, a universal spiritual fluid that crosses and regulates nature. Logos remains a question, an interrogation about a mystery revealed in the end times and the figure of the Spirit that transpires both as *Kosmos* and *Logos* in consciousness, that is, as an Intelligence in the sense of Anaxagoras, the principle of the order of things.

The research carried out by Hegel and Nietzsche on consciousness, as well as Heidegger's study of *Dasein* are precious not only for understanding the process of emergence linked to experience throughout time from a subjective point of view, but also for making *Telos* appear as a guiding principle so that completion can be achieved (with the mind and consciousness as a result, or for Heidegger, as a situation that has to be solved by means of anticipatory resolution, or Parousia). In the analysis of *Dasein*, the purpose is to go beyond *Telos* and to question Being, that is, *Kosmos*. It is, at this level, that the crucial point of philosophy's fundamental question about time is placed. *Dasein* delivered to death can be understood from the perspective of biographical time, and the question of Being presupposes the access to cosmological time. Here, we may have found another definition of the ontological difference between Being and the human being, which echoes the temporal difference between cosmological and biographical time. These two times are analytically separable, but are nonetheless connected. There is cosmos in our lives!

These considerations deserve a complete book so as to explain this cosmological and ontological (even theological) revolution of time that we have announced. We could discover that the access to cosmological time starting from biographical time belongs to the sphere of Mystery.

10

The Arrows of Time and Emergence

10.1. The three categories of time and contemporary physics

Contemporary science opposes two types of time, that of rational mechanics and that of thermodynamics, with the arrow of time leading to an increase in entropy. The reality of time is really contrasted. In other chapters, I drew a distinction between the time of the communicating scene and the time of the actors on stage [DUG 17]. The time of the scene appears in restrained relativity and is spatialized. We can even say that it is nothing and that nothing happens. What counts is the biographical time of the actors, and in the universe, the cosmological time of the stars and the masses that gravitate around one another. And which accompanies lunation, day and night, the seasons and the return of Halley's comet. This time seems cyclical, but the cycles are those of the masses that gravitate around the scene. The most important thing to bear in mind is that order is associated with time in such a way that cosmology is determined by *Kosmos*, interpreted as order of the natural or universal cosmos, or as the time that orders the emerging cosmos. At present, all the processes and material phenomena that we scientifically study are related to the physics or biology of emergence, which follow the logic of biographical time. Even if stationary states are observed, everything is transformed, modified and experienced! At the core of matter that transforms and stages itself via communication, the effects of the other two temporal modalities, *Kronos* and *Telos*, can also be detected.

The "physics of emergence" differentiates the dynamics of communication and the mechanics of arrangement, which nonetheless need to be articulated. As regards matter, communication influences arrangement, which would force us to admit that everything in nature is information-related and that the difference between things and beings is based on the organization of information. Mass contains information. Bosons convey it by means of material elements. In fact, there is a controlled transmission of bosons between the field and the dynamics of expression and reception placed at the boundaries between matter and "fields". Our cosmos with its masses and fields is an emergent and orderly system. However, matter favors emergences to a great extent because it has varying degrees of freedom and changes through communication. Matter is transformed because it is informed. Matter emits signals that may be either noise or information. This leads us to devise three categories of evolution and, following this, to three arrows of time, which accompany statistical physical processes in the following three directions: emerging order, disorder and stationary process (emergent order). For convention, the arrow of emerging order will be defined as negative, that of the noise as positive and that of stationary emerged order as neutral.

↑ Noise and emergence distortion

→ Time arrow → Structural stability and emergence preservation

↓ Information and creation of new emergences

Figure 10.1. *Noise, information and emergence*

In nature, these three types of processes compete against each other, at the time that they complement one another, within a unique composite system. If noise prevails, then emergent order partially collapses. If constructive information prevails, then a new emergence takes place, most often as a combination with the previous one. Constructive information results from communications captured and inserted by the system in a coherent manner. Thus, the components integrate or "semanticize" information by ordering action in accordance with a common purpose (because it has been transmitted) and determined by the principle of *Telos*. Communication within an organized system constitutes a transmission (biologists speak of "cellular transduction").

The diagram of three arrows is universal. It applies as much to systems described by non-equilibrium thermodynamics as to the living environment, with the emergence of life, then the species and finally, it applies to human societies, which are also determined by critical periods of social noise, language disorder or other moments marked by new forms, mainly in religion, philosophical thought, science and the arts, with the emergence of a new culture. While this scheme can be easily applied to the study of societies, it also makes it possible to describe the evolution of living environments with their species. Can this diagram provide a convenient explanation for the root causes of emergence, including statistical physics and the unveiling of the three arrows: (1) neutrality and stabilization of emergence with coherence in communications; (2) positivity and decrease (or change) of order with noise and (3) negativity and emergence of a new order with the integration of new information?

The principle of the three qualities (orientation) of time allows us to revisit the interpretation of the second principle of thermodynamics, by proposing a more universal principle governing all the possible transformations, and in particular, those of open systems, exchanging energy and information, and which are often appreciated in nature. At the most basic level of matter, information is carried by the elements of the "fermion-mass" or the "boson-field" class. However, in complex ensembles, fermion elements take part in a form of communication by acting as material support for information. This is the case of a hormone in the body (an ion attaching itself to the cell membrane), or of peripheral electrons at the material scale, ensuring the propagation of electric current in a conductive metal wire.

The three qualities of time determine at least the two classical categories of physical mechanisms. The neutral arrow corresponds to the Lagrangian formulation and to the principle of least action, which is also a principle of lesser communication, which is consistent with the idea of the structural stability of emergence. This arrow also explains how the emergent cosmos is described by Einstein tensors. The other two arrows describe the possible evolution of a system in which the order of communications increases or decreases depending on the effect produced by communications, whether they produce information or noise. The second law of thermodynamics is not a universal principle and only concerns the evolution of closed systems in which noise disrupts or destroys the original order.

Cosmology incorporates mathematical time used in two ways: in restricted relativity and in relativistic cosmology. The relationship of things to time is not the same and, furthermore, it does not describe the same things. In the first case, time is spatialized; it measures the interval between broadcast and reception between two things, separated by space. It is a time associated with communications. Cosmological time is that of arrangement. It does not elapse, but accompanies the change in the arrangement of things in space. Neither the photon nor the electromagnetic field produce time, except when the quanta of the field are absorbed by matter and later transformed into mechanical energy on planet Earth, from which derive the observed movements that make us aware of biographical time. This time is rooted in a cosmological time, and it is this point that remains the great enigma for the physics of the 21st Century.

10.2. Noise, temperature, entropy, time arrow

Temperature is a primitive notion related to the perception of hot and cold. In physical science, it is an intensive variable. If we add one cubic meter of air at temperature T to another cubic meter of air having the same temperature, then the two cubic meters have temperature T, and not 2T. Conversely, mass is an extensive variable. A kilogram of lead and a kilogram of iron placed on a scale weigh two kilograms. Temperature is measured with a thermometer. The operation seems quite simple. The principle of the thermometer is based on the proportionality between the expansion of a liquid and the increase in temperature. For instance, we use mercury in a vial connected to a tiny column that rises when the temperature increases. It suffices to leave the thermometer long enough to achieve equilibrium. If this is the case, then the heat flow from the liquid to the thermometer must be equal to the reverse heat flow. Thermal equilibrium does not mean an absence of heat exchange, but equality between two opposite thermal flows. The quantification of temperature has an arbitrary character. The Celsius scale is established by fixing the freezing point of water at 0 degrees and its boiling point at 100 degrees. Nevertheless, other scales, such as the Fahrenheit scale, can be used as well.

Temperature is a fundamental variable in thermodynamics as well as in statistical mechanics. At the microscopic level, it is not known how temperature is transmitted, but it is known that there are mechanical effects associated with temperature. We can mention, for example, the dilation of the liquid in the thermometer or the increase in the pressure of a balloon inflated with air when heated. The perfect gas equation describes this phenomenon: $PV = n.k.T$. This formula links a mechanical property, pressure, to temperature and the density of the present molecules: $P = n.k.T/V$. Boltzmann's constant k then appears as a "mechanical quantum", whereas Planck's constant is a "quantum of information".

Temperature takes part in a second context, that of the transformations that occur within a system. A new variable is introduced, entropy, S, which is an extensive variable. This quantity remains enigmatic, and scientists do not agree on its physical meaning, except for the idea of the disorder, which it is supposed to quantify. However, disorder, as well as order, is a relative notion. Entropy can only be understood in relation to temperature. A fundamental equation of thermodynamics associates temperature with entropy: $\Delta Q = T\Delta S$, where Q denotes heat and therefore ΔQ represents a heat exchange, which is proportional to temperature T and entropy variation ΔS. We can reverse the order of things with the formula: $\Delta S = \Delta Q/T$, where T quantifies a "thermal current" produced by systems that have an ambiguous link with mechanics, as evidenced by Brownian motion theorized in one of Einstein's articles published in 1905. This current is supposed to be symmetrical for almost balanced systems and can be reversibly transformed. These points were the object of endless controversy, not to mention the debate around the significance of entropy. The dispute is far from over.

In statistical physics, entropy is a quantity calculated using the formula $S = k.\log(W)$, in which k is Boltzmann's constant, the same used for perfect gases, and W is the number of microstates that can occupy the observed system in a laboratory. It is in this sense that entropy was interpreted as a deficit of information concerning microstates. In the context of the emergence of communications, I interpret entropy as noise. The equation of heat, $\Delta Q = T\Delta S$, describes a "noise" flow, where Q has the dimension of energy and is therefore equivalent to information, which we will consider as free. Hence a statement: heat is noise that corresponds to free information (untied, loosened).

For an irreversible transformation, $\Delta S > \Delta Q/T$. If this system is closed and does not exchange heat, then $\Delta Q = 0$ and we encounter the fundamental statement $\Delta S > 0$. Entropy can only increase; it has a physical meaning with the heat-insulating potential of a fluid, but it struggles to find a meaning in terms of quantification of disorder. We observe that constant k depends on the choice of temperature. If, by convention, we set the number of degrees between boiling and solidification of water at 1000 degrees (instead of 100), constant k is divided by 10 and, consequently, entropy is calculated by the formula $S = k.\log(W)$, also divided by 10. Nevertheless, TS quantity or $T\Delta S$ does not change. Heat flow does not depend on temperature conventions, which is reassuring for physical science.

Thus, TS quantity does not only denote the amount of heat-insulating power but also the energy lost by a system, lost in the sense that it does not take part in the transformations. It is for this reason that Helmholtz and Gibbs invented free energy F and free enthalpy H so as to quantify the capacity of a system to produce work or to transform itself. $F = U - TS$; $G = H - TS$; G function is decisive in thermo-chemistry. It is not a question of mechanical work, but it is a chemical reaction. Within the framework of the physical philosophy of emergences that I sketch, G indicates the communication potential that can be exchanged between the components of a system, assuming that not every "communication" is necessarily exchangeable. Indeed, there is "information" that is exchanged and wields transforming power, whereas "noise" corresponds to heat, to dispersion.

10.3. The effect of temperature on emerging order

In the majority of transitions of condensed matter, a decrease in temperature is associated with the emergence of material properties, whereas an increase in temperature makes these same properties disappear. Besides, this principle is commonly applied in chemistry. In order to crystallize a chemical substance, it is necessary to solubilize it and then put the solute in a refrigerator and wait, for a day or for a year. The crystal is an emergent order that occurs when communication between molecules takes place and ends up regulated in a coherent manner, which engenders a periodic spatial order.

Ferromagnetism is a phenomenon that depends on temperature. When a ferromagnetic material is placed in an external magnetic field, the moments of each atom acquire the same orientation that they partially retain when the field is removed. The magnetic moment is due to the spin, which is a quantum property that characterizes the communication of matter. At a microphysical scale, ferromagnetism can be partly explained by an exchange interaction, independently discovered by Heisenberg and Dirac. This interaction is based on Pauli's principle and the properties of the wave functions. We can interpret this as means of communication adopted by the electrons placed on (orbital) energetic layers. The exchange interaction increases or decreases the energy of the electrons depending on the superposition degree of the orbitals. This obviously reminds us about the considerations on quantum chemistry. The bonds between atoms also depend on the structure of the orbitals.

Ferromagnetism disappears as from a certain temperature threshold. The material then becomes paramagnetic. The phase transition is triggered at Curie's temperature characteristic of each material. The explanation of the phenomenon is apparently simple.

Thermal agitation destroys magnetic order. The material no longer retains its magnetization, but loses its memory. In ferromagnetism, this results in the hysteresis cycle. Therefore, heat produces thermal noise, which destroys the coherence of interatomic communications, at the time that it also destroys the physical (but fragile) memory of the material. There must be a link between communication, noise and memorization. Noise might alter not only the coherent communication of information but also memory understood as the result of information having influenced the "materiality" that received it.

Superfluid helium and superconductivity are also emerging properties related to temperature, which makes this order of matter disappear if it is too high. However, do not we observe an equivalent situation when, for example, heated liquid water becomes gaseous? Temperature produces noise and disturbs elementary communications in such a way that the fundamental properties of the liquid phase disappear. The "intermolecular bonds", which are responsible for the liquid phase, get broken by the effect of heat. Water molecules disperse in order to occupy a larger space, and get in order in a so-called gas phase. These observations add nothing to the description of things, but provide additional meaning by considering connections as the consequence of the communications established by "matter" at the microscopic level, whereas temperature corresponds to the quantification of the informational noise that takes place at the level of matter's communicating interfaces.

Fermi's theory of liquids explains how material communications lead to a coherent order, starting from a fermion gas, that is to say, from a configuration in which the particles do not communicate. Then, it is necessary to describe what happens when fermions are delicately placed in a communicational situation. This operation makes it possible to carry out an adiabatic transition (meaning there is no heat exchange), from the non-communicating phase to the communicating phase. This leads to the formation of quasi-particles. The mass of fermions is changed. Fermions are described as "dressed up", which actually means "surrounded". As if indifferent individuals located on a square started forming groups, or even a crowd. It is the interaction with the environment, in other words, quantum communication that provokes the transition from individuated fermion to the dressed up fermion, described as a quasi-particle. This transition is interpreted as a transformation in the structure of information when the fermions pass from the gaseous phase to the liquid phase.

During the liquid–gas transition, generated by fermion communication, the communicating properties (spin, charge, moment) do not vary, but the dynamic properties, such as mass, change and become renormalized. Excitation in gas and liquid correspond to one another, but in the second case, we have quasi-particles, that is to say, we have a kind of matter created by quantum glue when communications are established. It is like a set of guitar strings. They vibrate while ignoring each other, but if we place them in a situation where they are receptive and can communicate, then they vibrate coherently; they are quasi-strings, they sound louder, in the same way as (renormalized) quasi-particles endowed with larger mass do.

What about time in this description? Again, we have to force the interpretation. In adiabatic transition, there is no heat exchange, unlike the classical liquid–gas transition of water vapor, in which clouds encounter a mass of cold air, and by means of heat exchange, this cools down and then becomes liquid, thus generating rain. Fermion transition does not rely upon heat exchange. There is no informational noise entering or exiting. However, fermions must slowly be put in contact. In pictorial terms, the system must be placed out of disturbance and on the path of *Telos*. Finally, we will obtain quasi-particles of infinite duration. This is the sign that time related to *Kosmos* becomes efficient.

Fermi's theory of liquids is more important than we think because it indicates how fermion-like particles such as electrons can associate in order to construe communicating structures. Not only bosons communicate in the material world. Everything happens as if a spider's web was formed from elementary components capable of missing their individuality (deletion by the principle of *Telos*), so as to blend into the canvas and enable remote communications. This observation leads to a universal physical principle: *every ordered phase of matter depends on a specific mode of communication between the components*. Temperature has an influence on emergence because it is the measure of the noise that is opposed to order in the circulation and sharing of information between the system's components.

10.4. Order and the arrow of time

10.4.1. *Entropy and disorder*

The arrow of time has been poorly understood because so far it has been applied to heat exchanges and, therefore, to noise generation. However, the arrow of time is also oriented toward order and emergence. For this, it is necessary to study the arrow within situations where it is not the noise, but the shared information that prevails and generates emergence from the components, all of this described as both an arrangement structure and overall coherence and adjustments in communication. The comprehension of an emerging order will still be a challenge for 21st-Century scientists. The approach toward complexity has been misleading for more than a century because of a hasty interpretation of thermodynamics and of the state variable entropy with a negentropy that makes no sense, especially when applied to living systems. I must reiterate that order is not the opposite of disorder. There is no symmetry between order and disorder. Order is an emerging quality of the system.

1) In the first place, physics has obscured the difference between arrangements and communications. Despite some scholarly developments, entropy was not understood. Jacques Tonnelat drew a distinction between spatial complexity and energetic complexity so as to describe "thermodynamic" systems. While the first one is divided into spatial distribution and conformation complexity, the second one is divided into particle distribution depending on energy levels and binding complexity. This very subtle analysis reveals the distinction between arrangement and communication as well as the complex interaction between these two poles [TON 91, p. 77].

2) Another source of confusion comes from the use of the notion of entropy to understand the living world. Order is interpreted as the opposite of disorder, in other words as an entropy of opposite sign, known as negentropy. In fact, if we admit that disorder produces

entropy, then order cannot be conceived with an entropy diminution. There is no justification to symmetrically deal with disorder and order. The systemics of the 20th Century played with this symmetry by resorting to the notion of information. Hence, we get the principle of order generation by noise. Now, noise can only create disorder and alter emergences. Only integrated information can create order. Noise then represents a type of information that does not fit into organized order and even disrupts it. The universe (cosmos) is determined by order, whereas in our world, disorder is not only present but also necessary for paving the way to transformation and evolution processes, conducive to emergences associated with biographical time. Noise is only the consequence of the free production of information in material ensembles. It is the condition for unintentionally created emergences via material experiences and communications.

The arrow of time accompanies the generation of noise, of not only disorder but also order. As much as the transformation toward disorder is easy to understand, the processes of emergence are difficult to capture because they are based on a set of conditions that analysis can barely elucidate. Among these conditions, a structural threshold in communications must be achieved. The components share their experiences and eventually find a rule for ordering their operations. It is the principle of *Telos* that becomes efficient. The procedures leading to emergence take time. Systemicians employed the image of von Foerster's cubes as a metaphor for self-organization. These magnetized cubes are placed in a container that is shaken up, where they associate in order to create composite structures. This image is too mechanistic and atomistic because it considers the components as individuated objects, endowed with specific properties. However, the physics of emergence is radically different. Even if they exist, objects do not yet have complete systemic properties. They must interact, communicate and exchange so as to determine themselves and find the common way to emergence. In other words, taking up von Foerster's image, cubes do not have magnetization, they must create it and communicate it to the other cubes before any systemic combination can take place. Therefore, components must be individuated and differentiate themselves, so that *Telos* can later erase the expression of certain properties in order to merge into the whole of the system. In the physics of emergence, components are not pre-established objects, but are modified along with the system's transformation.

10.4.2. *Prigogine and irreversibility*

Irreversibility was a primitive or, shall we say, a rather trivial notion for the ancients. A broken ceramic block does not recover. Living creatures inexorably tend to degrade and then die, and do not come back to life. It was in the 19th Century that irreversibility entered the field of mathematical physics, thanks to the introduction of entropy. This situation provoked one of two great dilemmas of the 20th Century (the other being the classical quantum schism). How can we understand irreversibility while the laws of classical and celestial mechanics are symmetrical and deterministic in principle? Prigogine points out that Newton's laws dominated theoretical physics for three centuries. In the introduction to his essay *The End of Certainty*, written as a testament, he reformulated the question that haunted physicists in the 20th century: how could it be possible for the arrow of time to emerge from a world to which physics attributes temporal symmetry? [PRI 96, p. 10]

The wording of this question is not neutral. It assumes that the Newtonian world is more fundamental than the world of irreversibility. The physics of emergence invites us to reverse the question and to wonder how Newton's deterministic world can emerge from a world determined by chaos and how unlimited exchanges of information and the arrow of time (with its two complementary poles, *Kronos* and *Telos*) are conducive to the third pole, that of *Kosmos*. In other words, how *Kosmos,* with its cosmic clock, can emerge from *Kronos*. In the beginning, there was the quantum! This eventuality was examined by Prigogine, who concluded his essay by imagining a universe resulting from primordial instability, and whose genesis could be described by irreversibility and non-equilibrium physics. However, there is still the possibility of having no explanation to the origin of the universe. Even if it were so, this does not prevent us from questioning the link between matter, the cosmos and time.

Prigogine's book is what we call an era marker, as was the case of Monod, published in 1970 with the consequence of science being anchored in atomism and mechanism. Prigogine tried to pave his way toward a new world, that of instability and of creation in nature. The exit of Newtonian physics was guided by quantum mechanics as well as statistical physics. As noted by Prigogine, the notion of trajectory was no longer used in the descriptions of interacting systems governed by "statistical laws". Quantum physics also kept its distance from space, in not only the formulation of state vectors but also the description of two particles, which made it necessary to split each space coordinate, and therefore to configure a description with six parameters. In the two physics, the notion of trajectory is no longer suitable, but for different reasons. This is the first lesson we can draw from the postmodern conception of matter. Only a physics of arrangements can content itself with a description taking trajectories into consideration.

Interaction modes in irreversible systems only acquire physical meaning if we consider particles as components that "plunge" in an environment where interactions are persistent, whereas classical mechanics applies to situations where interactions are transitory, hiding the assumption that movements are isolated. The distinction between transitory and persistent interactions requires a distinction between "localized" and "delocalized" probability distributions [PRI 96, p. 133]. The unification of dynamics with thermodynamics is impossible with trajectories, but with the new formulation, this objective becomes accessible. Long-range correlations appear when the "forces" between molecules are of a short range. Prigogine's formulation describes the mesoscopic emergence of correlations starting from the microscopic level of matter particles.

Then, the system behaves as a field whose excitations are given by global and coherent dynamics describing the coordinated processes of the particles. These correlations characterize the dissipative structures that emerge like quasi-particles in Fermi liquids. Nevertheless, these two types of transition do not occur under the same conditions. There is no heat exchange in Fermi's liquid–gas transition, contrary to what occurs in thermodynamics. Besides, time is considered as the lifespan of quasi-particles and placed as a result of the Fermi–Landau statistical rules, whereas in Prigogine's description, the time associated with the persistence of "resonances" is a necessary condition for revealing emergent structures. This specific time of statistical thermodynamics has *Kronos* as a principle, as an affirmation, as expression. The Fermi–Landau transition occurs as if the fermions were tiny vibrating strings that are delicately communicated in the context of thermal silence and, in the end, an extended string vibrated: the quasi-particle. Prigogine's transition can be compared to a huge orchestra that plays musical chaos, where each instrument plays its score, listens to other instruments and then borrows other scores. From this chaotic hubbub, a common score (something that we call "correlations") will eventually emerge.

According to Prigogine, the resonances in dynamics contribute to communication, by putting the field frequencies and the particle frequencies in relation. In the end, this remote order can be interpreted as a description in terms of the physics of communications. It is as if the system helped communication routes to emerge, in a similar manner to a set of strings rubbing against each other (collisions and diffusions), and finally vibrating in such a way that a vibration is transmitted from one place to another (correlations). In physical terms, these correlations make it possible to lower the energy of the system, which is interpreted as the efficiency of the *Telos* pole moderating expressions with suppressive influence. Nevertheless, the image of the strings is misleading because the theoretical description does not indicate the location of the strings but rather the way in which they vibrate, that is to say, a distribution of frequencies or even better, a score where the notes of the vibrating matter produces shapeless musical chaos or suggests some kind of formal music.

We thus measure the "genealogical abyss" between *the artificial transmission systems* studied by Shannon and *natural communication* in matter and in life, whose principle is the joint emergence of communication systems and of the signals that go through them. Then, the arrow of time is interpreted as a qualitative asymmetry between a system that "communicates" with noise and a self-constituted system that exchanges signals and not noise. This supposes the emergence of coherence between emitted signals and receivers.

As a conclusion, there is a close link between irreversibility, communication and emergence. If the world of communication corresponds to the *modes* of Spinoza's substance, then the *divine attributes* of this same substance can be interpreted as "arranged information". However, the Spinozian scheme is no longer suitable because *modes* are capable of generating (submerged) *attributes*.

10.5. From *Kronos* to *Telos*, irreversibility and the two arrows of material time

10.5.1. *The RHS from the Brussels-Austin school*

The origin of irreversibility is still unresolved. This was the conclusion formulated by Prigogine with his colleagues from the Brussels-Austin school [BIS 03]. In fact, there are two stages in the research and conceptions developed around Prigogine. One was achieved in the late 1970s with two books, *The New Covenant*, co-written by Isabelle Stengers and *Between Time and Eternity*. The second stage is summarized in the book *The End of Certainty*. While both approaches use quantum mechanics, they differ in the tools employed. The first approach resorts to a conventional formulation with Hilbert spaces, while the second one, less familiar, uses a tool derived from the first, the *rigged Hilbert space*, known as RHS. Nevertheless, "the physical reason that makes it possible to explain why the arrow of time goes from the past to the future is not clear with the RHS approach, although this approach offers us a mathematical tool for exploring and describing the arrow with further details" [BIS 03, p. 10]. The quantum and statistical formulation of thermodynamics is as problematic as early quantum mechanics was. It entices us to reason using mathematical tools, which, although unable to deliver the solution, can certainly shed some light on the question.

Poincaré observed that systems showing resonances between degrees of freedom are not integrable. These resonances possess physical meaning: they describe the opening of a communication channel for transferring energy (specified by a frequency) from one physical component to another. In the case of a system with two degrees of freedom, the trajectory in the state space is indicated by the Hamiltonian. Now, let us add a small perturbation, which makes the Hamiltonian unintegrable (situation with resonance). The trajectories set in the space between states take two forms, namely an elliptic orbit, which corresponds to a stationary orbit, and an erratic orbit, which is sensitive to initial conditions and produces deterministic chaos with its strange attractors. At this phase, we are still in the domain of classical mechanics. Later on, Prigogine and his colleagues described a situation in which the number of degrees of freedom is large, even infinite. This situation corresponds to physical realities, such as the interaction of an electromagnetic oscillator with a field with infinite degrees of freedom and many other phenomena that belong to the sphere of quantum mechanics or statistical physics.

The RHS space is then used for solving the equations formulated in order to describe these large Poincaré ensembles with their resonances (communication). The Brussels-Austin method comprises three procedures: (1) dynamic description with a distribution function including resonances, (2) RHS space as a tool for solving equations and (3) an appropriate time for putting dynamics in order [BIS 03].

1) Prigogine and his colleagues justified a description with *distributions* because irreversibility cannot be deduced from a representation with *trajectories* (and therefore in the language of the physics of arrangements). On the contrary, irreversibility arises from a global systemic configuration linked to the persistent interaction of components, that is to say, resonances and, therefore, with energy, which is communicated from one component to another. The persistence of resonances reflects the extended duration of energy transfers, and this idea is interpreted as a shift from the physics of arrangements to a physics of communications. This is useful for describing transformation processes related to the arrow of time. We can thus appreciate the abyssal difference that exists between this and the trivial irreversibility of the vase breaking into a thousand pieces.

Interactions producing breakage are limited in number, and above all, they only last for a moment. Contrary to this, resonances in thermodynamics can almost last indefinitely.

2) The RHS space is a mathematical tool derived from Hilbert's space (HS) used in quantum physics for describing the preparation of a system in the form of state vectors, a specific topology, vector norms and Dirac's bra and ket notation system. The RHS space is built upon the previous one, by introducing a set of norms that depend on operators and observables. Unlike the HS space, which is universal, the RHS space is singular and its choice depends on the system that we prepare and observe. Two important things must be pointed out. First, there is a similarity of the RHS approach with the tensor networks used in the physics of condensed matter, and whose choice is also adapted to the physical situation described above. There is another similarity though: the RHS approach uses a reformatted Hilbert space, and the tensor networks result in a description of the system at a corner of Hilbert's space. The other important point is information. If we admit that Hilbert's space describes fundamental quantum information, then the RHS procedure introduces a topology of information, just as tensor networks do. From this, an interpretation in terms of the informational order of matter can be derived, an order that reveals submerged structures.

3) The third aspect is certainly the most enigmatic one. As Bishop observed, the Brussels-Austin approach sheds light on the arrow of time but does not explain it. Antoniou and Prigogine introduced a prior temporal condition. The system's excitations and preparations are interpreted as events occurring between a −t and a 0 time, while de-excitations and detections are interpreted as events taking place between 0 and +t [BIS 03, p.10]. Therefore, the arrow of time is not the emerging consequence of the dynamic description because it has already been introduced in the premises of the argument. We observe a type of circularity similar to the one found in quantum Darwinism, which, according to Ruth Kastner, points to a classical world, because classical elements are introduced from the beginning (see Chapter 4). It is necessary to leave this frame of thought.

10.5.2. *The two orientations of time and the metamorphosis of modern physics*

The RHS approach shows two categories of evolution. The first determines the succession of states in a sequence oriented from the past to the future, and finally, a state is obtained when equilibrium is reached, thanks to correlations. The second operator works in the opposite direction and describes systems whose balance was achieved in the past. Now, as we have never observed a system going toward the past, this operator is considered non-physical, unlike the first one, which describes the irreversible evolution toward equilibrium in the future. However, this conclusion is not definitive. It is possible to find a physical meaning for evolution moving from the future to the past. This is what Bishop suggested in a study published some 10 years ago, and which has remained rather confidential because the ideas that are exposed on it transgress the self-imposed limitations of Newton-inspired modern science, particularly, the exclusion any sort of finalism in nature. However, the "thermostat" evolution toward the past might also be linked to final causes [BIS 05].

Nevertheless, I will not follow Bishop's subtle conclusions based on the duality between the two dynamics, one toward the future linked to matter and the other toward the past, which might be linked to the *mind*. I would opt for a "subject matter pole" that receives information related to an "object matter pole". The subject-attribute returns to the stage which might suggest a redemption of the modern physics born with Galileo and Newton, involving the principle of the retraction in the human subject; in other words, this would imply setting aside the subject matter that represents the receptive aspect of nature. Matter expresses itself, communicates, receives and sometimes informs itself. It is this novel scheme that emerges with quantum mechanics, especially when it is used for describing the evolution of systems with a large numbers of particles, with their persistent resonances.

To put it in terms of natural ontology, the two orientations (*forward* and *backward*) resulting from the RHS approach in the mechanics of irreversible processes are at the basis of irreversibility and the arrow of time that we know. The arrow of time is then doubly determined, with dynamics pointing toward the future and the past.

These two dynamics determine the expression of phenomena and their reception, with a regulated articulation. The arrow of time is simultaneously determined by the expressed matter pole and by receptive matter. It can be compared to a liquid whose shape is adjusted for filling a vase. This is the reason why I do not agree with Bishop's interpretation on the role of the mind. We are not the authors of the arrow of time, but its children! On the contrary, linking purpose to orientation toward the past is a good idea, provided that teleology is reserved for not only man but also the whole of nature. The dynamics of matter toward the past cannot be interpreted in terms of the mind because it is at the same level as the future, and is not hierarchical. If emergence and purpose are present in matter, then they are horizontal, not vertical.

If we go back to matter, the double determination of time refers to the game between *Kronos* and *Telos*. *Kronos* is headed to the future. *Telos* is also directed toward the future and finality, but it rests on the arrow directed toward the past, which, by reflective effect, turns around in a similar way to the reflexivity of the essence conceived by Hegel. This is also the idea suggested by Bishop, a reflexive reversal of the arrow directed toward the past and which enables the emergence of purposes (related to the pole of the *mind*). I will therefore insist on the idea of the double origin of the arrow of time, involving two tendencies: *Kronos* directed toward the future and *Telos* toward the past, memory and conservation. Matter-information causes the arrow of biographical as well as of cosmological time. *Telos* and *Kronos* are, respectively, related to anti-matter and matter, but as there is an excess in matter, the resulting arrow goes to the future. This arrow enables communication with both poles: expression and reception. Noise represents that which is formless, order reveals itself in the expressions accompanied by structures and responsible for macroscopic emergences, at our scale, or for mesoscopic emergences in the systems studied by statistical physics. As Prigogine had cleverly suggested, through the arrow of time and the non-equilibrium situation, matter "sees". Therefore, instability must be related to the communication of matter with resonances. The direction of time governed by *Telos* enables matter to receive signals and, if necessary, to inform itself. These considerations about the two directions can be summarized as shown in Figures 10.2 and 10.3.

↑ chaos

past ←← →→ **future**

↓ order

Figure 10.2. *Order and time*

↑ noise

Telos ←← →→ **Kronos**

↓ ← information →↓ purpose

Figure 10.3. *Information and time*

I conceive a dialectic between *Telos* and *Kronos* with two categories of processes, some interpretable as noise and others as information taking part in the emergence of a purpose whose theoretical result is translated by correlations and whose empirical result is reflected by observable structures. We are also witnessing the emergence of a field of communications linking the two poles of matter, expression and reception. The components of the non-equilibrium system communicate, as if they were quantum systems being able to extract information by producing observables. This suggestion makes sense with the RHS formulation developed from basic quantum theory and in which the χ_i observables generated from a Lie group are inserted for specifying the norm and therefore the structure of the available information related to vectors of the RHS space [BIS 05]. Let us observe that in the quantum field, Lie groups generate bosons that mediate interactions. Therefore, the observables of RHS formalism follow the same formal order as elementary bosons. This reinforces the idea of matter communicating with signals endowed with certain coherence, thanks to generating structures. The two universal directions of time have been unveiled by not only statistical physics but also the analysis of Heisenberg's equation in the framework of matrices quantum mechanics (see Chapter 2). In this sense, evanescent resonances (*decay*) resulting from statistical dynamics can be interpreted as informational processes (related to the mutual observation of components). Not only does matter express itself, but is also imprinted.

10.5.3. *The arrow of* Kosmos; *mechanics and thermodynamics*

Now we can go back to the question of the symmetrical time arrow, present in the equations of mechanics and of gravity. This arrow is based on an absence of dialectic between *Kronos* and *Telos*, or rather a perfect symmetry because there is neither noise nor finality. There is no emergence, but an emerging cosmos, which can be symbolized with this diagram, derived from the previous one. The order of *Kosmos* is deterministic, symmetrical by time reversal. It is possible to overthrow the classical ontology of the cosmos. What is most fundamental is the arrow of irreversible time conceived as dialectics between two poles: *Kronos* and *Telos*. In this context, the reversible arrow associated with *Kosmos* results from an emergence and dialectics resolved between two poles. The reversible arrow is deduced from the irreversible arrow, and not the other way around. Our world is an emerged cosmos, resulting from *Kronos* and *Telos*. The resolution of *Kosmos* is none other than universal gravity, implying not a biographical but a cosmological, somehow universal time.

Telos ←← { KOSMOS } →→ *Kronos*

Now, let us imagine that this cosmic order is disrupted, for example, picking up a stone and then throwing it in the air. There is an instant reaction, settled, without any dialectics. The movement of the stone is determined by the two poles of time, *Telos* and *Kronos*, which merge and become one. *Kronos* translates the movement and *Telos*, the natural destination fixed by *Kosmos*, employs Aristotle's terms. The mechanical alteration of the macroscopic world is regulated by classical and rational physics. Thermal alteration is the result of quantum and persistent communications. Let us recall the distinction we made concerning the vase that breaks into pieces, based on the principle of mechanical noise. The arrangement of matter is modified. With thermal noise, matter communicates and is transformed as correlations emerge. Mechanics and thermodynamics are irreconcilable because thermodynamics is based on the principle of quantum communication. Therefore, the approach to non-equilibrium physics is completely foreign to Newton's laws, as Prigogine pointed out, as well as to Einstein's cosmological laws. The schism between the two physics is also perceived through the duality between cosmological and biographical time. Finally, irreversibility and biographical time on our planet rest upon the quantum action of a very well-known source: the electromagnetic radiation coming from the sun.

Mechanics and thermodynamics are connected by means of thermo-mechanical interfaces. When we activate the breaks, the disks heat up under the effect of pad pressure. Mechanical noise then becomes thermal noise. The opposite also occurs when the thermal energy of fuel combustion is turned into mechanical energy, thanks to the combustion engine. The time of {*Kosmos*} is therefore the time that regulates the scene and the succession of arrangements with the coherence of perceived phenomena. *Kronos* and *Telos* determine the processes of quantum matter and communications (expression and reception with information). The only way to find a quantum formulation of cosmology is to take into account the two poles/time directions. That is to say, time elapses in the sense of the future, but another time elapses in the direction of the past. At present, we will stick to the idea of irreducibility between classical physics and the statistical formulation of non-equilibrium, especially in what concerns the RHS approach.

10.6. *Kosmos*, *Kronos* and *Telos*

Reflections on the plural orientation of time are associated with the figures of time inherited from antiquity. Nevertheless, these figures must be resumed and conceptualized in order to fit with the results of contemporary physics obtained in three branches, namely quantum dynamics, classical mechanics (with cosmology) and thermodynamics.

1) *Kronos* is the god who devours his children. In antiquity, it used to symbolize the aspect of time that destroys works. The principle of *Kronos* has now expanded, and can also refer to a positive direction of time, which facilitates the emergence of transformations. The second law, that is to say, the increase of entropy in a closed system, is derived from this. In non-equilibrium systems, *Kronos* sometimes produces communications leading to the emergence of natural orders. In both cases, *Telos* is also involved, but not in the same way. In a closed system, there is only evolution tending to equilibrium. In a non-equilibrium system, two dynamics are produced (as part of the RHS approach).

2) The principle of *Telos* extrapolated from the Aristotelian entelechy helps a system to evolve toward equilibrium accompanied by emergences. Teleology is key to any philosophy that presupposes that things are not "blind" and manage to orient themselves by benefiting from material contingencies. *Telos* is a reasonable hypothesis within the framework of the doctrine that connects emergencies with communications. If the parties "get along", so to speak, then they can manage to follow a collective direction and create emergences involving verticality, if possible, and a nucleus performing in a similar way to a musician-conducted orchestra. If mechanical physics leads to a philosophy of chance and necessity, paraphrasing Monod's ancient formula, then the physics of communications makes it possible to introduce *Telos* without any difficulty. And this, from the very stage of material emergences, leads to the origins of life with enlightened, oriented chance as its central idea.

The effect of *Kronos* is faster than that of *Telos*. Economists know that it is easier to destroy jobs than to create them. This common sense remark illustrates the conjecture of the two poles of time with *Telos*, whose effectiveness is not instantaneous. It takes time for emergences to occur, 1 to 2 billion years for the appearance of life. Hundreds of millions of years ago for the species, and regarding humankind, it took millions of years to pass from the anthropoid to the homo sapiens, then a few tens of thousands of years to reach the stages of the big human enterprises in China, Egypt and Assyria. *Telos* often implies a persistent orientation, and for humans, that is translated by perseverance. Thus, we reencounter the idea of *conatus* in Spinoza. The components focus on communicating, orienting, arranging themselves and memorizing their experience and end up finding order so as to create emergences. A central core emerges. Communicating structures emerge with signals that are exchanged and interpreted: molecular messengers, RNAs, synapses, language, etc.

Gould's studies uncovered sudden appearances of diverse and numerous species. These periods of creation of the living only reveal the effectiveness of the successful dialectics between *Telos* and *Kronos*, which disturbs the emerged (or, shall we say, balanced) cosmos of the previous species established in duration, by creating new forms. The appearance of human societies, then of large enterprises with cities, also reflects the effectiveness of *Telos* with the appearance of functional communication structures for organizing these large ensembles. The origins of life are also linked to *Telos*, which has become effective in mesoscopic emergence and enabled proteins and nucleic molecules to communicate and to select communicating structures such as RNAs.

The diagrams below show the principles of evolution with adaptation and resolution periods referring back to the quasi-balances of Gould and Eldredge (line 1), as well as periods in which the new species emerge, referring to equilibria "punctuations" (line 2). It is the principle of *Kronos* that creates an imbalance, which ultimately leads to the emergence of a new species, by working in tandem with the pole of *Telos*.

1) *Telos* of the emerged world ←← →→ *(equilibrium adaptation)*

2) *Telos* of emergences ←← →→ → *evolution, creation*

The following are information about *Kosmos*, a term that referred to the immutable order of the universe for the Pythagoreans. In this study, *Kosmos* represents an *arranged cosmic time* linked to classical matter that flows by creating the conditions for the actors to communicate on the stage and play out. *Kosmos* is the natural physical order on which the other two modalities of time resulting from communicating matter are superimposed. According to the circumstances, the "game" of *Kronos* and *Telos* produces either disorder and noise or order and structure. *Kronos* regulated by *Telos* engenders the unveiling of emergences related to submerged and veiled structures.

10.7. Concluding remark on physics and things

We might suggest that in the beginning there was quantum, accompanied by *Kronos*, the principle of action related to elementary quantum. This proposal opens up to a new and astonishing cosmogony: the fundamental constants, h and c, condition quantum communication at the level of the hardware interfaces and then in the immaterial extent. The third fundamental constant is G; but behind gravity, something more essential is hidden: mass. The reason for this is that G regulates arrangement in the cosmos, particularly the succession of arrangements governed by *Kosmos* cosmological time. However, while G is introduced as a universal constant, the same can be said about the masses of each particle of matter (protons, neutrons, electrons), which are also fundamental constants in our cosmos. This mass of material particles indicates the "composition" of quantum information, which enables the creation of coherent order in the universe. Particles are not objects but "blocks of basic information", welded by an organizing principle and, if necessary, able to communicate.

The argument we have explored so far has made it possible to clarify some points. Action and information are ontologically identical. The classical laws of physics and the laws of quantum dynamics use the Lagrangian. After Feynman and Susskind, quantum physics is the most fundamental science, and we now know that it is information that is fundamental. It is at the core of the principle of least action, which should rather be called the principle of less information, related to *Kosmos*. In the world of transformations, action and information work with the principle of "diffuse *Kronos*" and create the arrow of biographical time. Now I will introduce some general notes to summarize the still enigmatic question of time:

1) Cosmological time is based on unaltered information, that is to say, the principle of less information (or deformation), with a fixed resolution between the two orientations, *Kronos* and *Telos*, for a universe regulated by the *Kosmos*, in which the masses affirm themselves while they are regulated by inertia and gravity (Newton, Lagrange, Einstein) and, perhaps, by strong force. The masses express themselves on the extended stage by arranging themselves at every moment in time.

2) Biographical time is that of transformations, of information. Paraphrasing Leibniz, we could invoke a principle of information enough for describing the conditions of matter transformation, with two categories of evolution, one linked to noise and disorder, with an excessive *Kronos*, and the other tempered by the suppressive and moderating effect of *Telos*. This description does not solve the enigma of time, it poses it. The processes accompanied by biographical time rely on the electromagnetic interaction, which informs the systems. In complex ensembles, this interaction produces motion, thermal movements in inert matter as well as electromechanical movements in the self-propelled living. The components of an ordered system are governed by orientations. In a cell, in an organ, there is no component translation, but an assembly of components determined by orientations. If we note that there is a homomorphism between SU(2) and SO(3) groups, then we can think that the weak force determines orientations.

3) Phenomenological time is spatialized. It is the time of special relativity. Nothing happens, except for the propagation of basic "quanta of information".

As a conclusion, we will state that *Kronos* and *Telos* constitute the two polarities, the two orientations of the essence of time. These two essences are dialectically united by a large *Kosmos*, essence of the megascopic universe with cosmological time, and a small *Kosmos*, related to our macroscopic world and its biographical time. These time essences enable an adjustment of arrangements, that is to say, not only order in the universe, but also order for the action of the components in a system, for living creatures in their environment, as well as coherence in perceptions. This essence of time is related to Kant's *a priori* time, but is not reserved to the human subject; it is present in all forms of matter, from the mesoscopic to the megascopic, through the macroscopic.

If matter is equivalent to information, then, at the cosmological level, time will be interpreted as order in information, an order governed by this emergent property that is the gravity present in our cosmos. In our world of transformation, another time is emerging and leads to a set of emergences such as those of condensed matter or life. The following two hypotheses make it possible to conclude: (1) Quantum gravity is conceived as an order made up of hadronic matter and strong force, with two poles, matter and anti-matter, in relation to the two determinations of time. (2) When ordered, electromagnetic interaction is at the origin of the living. Magnetism is undoubtedly at the root of an order that is involved in quantum matter as well as in life. There is a principle governing life, in the same way that gravity governs the cosmos.

10.8. From the history of black holes to the light of time

The black hole is a cosmological object calculated from the equations of relativistic cosmology. In a strictly relativistic context, a black hole cannot radiate. Although it presents itself as an object regulated by the *Kosmos* principle, nothing can come out from a black hole. Its history is, so to speak, unfinished. We even wonder whether it has a background. To make the existence of this object true, physicists provided an explanation framed against biographical time. The black hole is supposed to have originated from an end-of-life mass star, in which the mass density became so high that it curved space-time in such a way that even a mass-free photon could not escape. Let us not forget the universal principle that was deduced from that. There is no biographical time without matter that communicates, creates noise and has a temperature, which is the case of the black hole theorized by Hawking, to which a temperature was attributed and whose quantum fluctuations were interpreted as communications without prejudging the place where they were supposed to take place.

Conclusion: *the classical black hole no longer has a history, and the quantum black hole still has one.*

From the ontological, or rather ontophysical, point of view, time is produced by quantum processes, which are no other than communications, leading to information, considered in the sense of modifications, about the influences on a system when this information is received or perceived. Time is produced by quantum information, but for this to be time, information must be coordinated, because time is a concept from the macroscopic world, whether apprehended by subjective consciousness or measured by modern science. This time expresses itself in the present, which is a subjective notion whose emergence presupposes certain duration and order in information. *Duration is the being of instimes*, that is to say, that which reunites instimes. In the science of physical things, time expresses itself through space, which is also an emergence. Space is extended time, an ensemble of apparent forms woven by a setting, which undoubtedly combines two global orders, namely magnetism and gravity.

It is usual in physics to distinguish two tenses, the one that elapses and the one that accompanies transformations, in other words, the (thermodynamic) arrow of time. The real time that elapses is what I call cosmological, even universal, time, while the one that transforms is biographical time, it accompanies the transformations of the systems caused by the communications and influential information. The separation between the two types of time is misleading because, even in thermodynamic processes, biographical time is intertwined with cosmological time, in the same way as the fact of being is intertwined with living creatures.

Cosmological time can be conceived as that which emerges from an information set intended to be compressed and decompressed. I mentioned the expression and its complement, impression. However, we must also consider the compression of information (related to memory and duration). This possibility can be explained with the metaphor of the music score, with its staves and notes. The staves correspond to the time, punctuated by the metronome, whereas the notes correspond to information (frequency and duration). In this metaphor, the notes are inscribed in time, which is external. In the cosmos, time emerges from ordered information. It is as if the musical notes were placed in the right order so that staves can emerge from this succession.

Matter is composed of compressed quantum information, destined to be expressed. The parameters of quantum theory are set according to the time of special relativity, which is that of communication. Time and space make it possible to register an infrascopic system (particle, information, state vector). However, for two systems, it is necessary to have another space-time. In a multi-particle system, space is not a physical space. As a result, time in the macroscopic sense does not emerge from the quantum world. Time only makes it possible to inscribe the appearance of a phenomenon in quantum observation. And from this, we come to the idea of phenomenological time as a condition for phenomena to meet in space. This time makes communications possible.

Time also accompanies an order in the succession of phenomena on the stage, which enables meetings, communications, the emergence of structures and the possibility of weaving common stories. Time connects things by making communication possible. Thanks to these meetings, systems are informed, and, if necessary, emergences take place. Every emergence has a story and occurs within a story. The submerged nucleus represents the "compressed" result of this story, that is to say, the biographical being set against a temporality, which is also biographical. However, we must not forget the cosmological being with elapsing time, which maintains the scene of the cosmos and makes the advent of structures possible, as well as the realization of stories by the different actors of the scene: cells, plants, animals and men.

Bergson used to think that the universe is a god-creating machine. In fact, the universe is a scene on which individuals with a "structural vehicle", the body, emerge, enabling them to engage in nature so that they can become involved in a story, or even invent stories, as man has done from the moment he could use speech. *The universe, a story-producing machine*! If language has enabled man to shape the great cities in the times of Sumer, by enabling him to speak the language of enterprise and order, according to Spengler, it has also made the elaboration of myths possible, which are stories, essentially. Essential time makes the development of stories possible, be them quantum matter, the thermodynamic arrow of time or the "scientific myth" of the black hole, including modern man and his narrations, from the first cells to animal species and so on.

To sum up, the time of phenomenology is spatialized and relates things and beings with their stories. Biographical time (the double orientation) is that of transformations, changes and metamorphoses, and finally, cosmological time is that of the block-universe, of the mechanics of arranged masses. The dialectics between *Kronos* and *Telos* does not stop in biographical time, because stories are open, some leading consciousness toward the *Kairos* (illumination, advent, revelation). In the block-universe, the dialectics between *Kronos* and *Telos* is solved (*Kosmos*); this constitutes the scene of the cosmos with its material and spatial configurations (arrangement), succeeding one another. Biographical beings play on this stage, invent themselves, transform themselves, have stories and create stories. This triplet was foreshadowed in a study in which three domains were conceived, namely expression-reception, "process" and incarnation [DUG 98]. There, I developed *the concept of Time/Mirror conversion, which anticipates the thesis of compressed information, whose mirror structure is undoubted, if we consider the universal symmetry of the quantum world (CPT among others).*

The following are information about quantum gravity, whose essential problem lies in the specification of the two times in physics, one setting the parameters for quantum phenomenological processes while accompanying the biographical processes of transformation, and the other putting arranged things in ontological relation with cosmological time. From the viewpoint of physical philosophy, quantum gravity consists of bringing phenomenological time and the time of being together. In the middle are the biographical time of beings and of trials. *Dasein* then represents the embodied resolution of quantum gravity. How is it possible to offer a mathematical theory? Physicists are holding the cards in their hands so as to find the solution, if such thing exists.

Finally yet importantly, we will remember the idea of "compressed time", while on the stage, actors play their story, express themselves and inform themselves with the rules of time that expresses itself as if it were decompressing, to dialectically recompress itself. This process involves both polarities, *Kronos* and *Telos*. But strictly speaking, it is not legitimate to essentialize time. This makes it necessary to formulate the idea differently, by suggesting that it is information that is compressed and which decompresses itself, produces two times, that of the elapsing scene and the one that unfolds, thanks to the actors of the scene. Thus, the compressed information of the megascopic universe is at the source of cosmological time. Biographical time is produced by the decompressed macroscopic biographical information, the one that constitutes psychoanalytic material for humans, who are historical creatures after all. DNA is then interpreted as mesoscopic information related to the temporal life of cells, in a medium or an organism. Finally, let us recall the infrascopic information of the quantum atom (see Chapter 2 section 5). Chapter 11 will be devoted to the different "scopic" scales.

Time is thus conceived as compressed and ordered information. Quantum physics only describes communications and elementary information compressed in the space of vectors and basic states. It has made it possible to trace a parallelism between quantum matter and information. The inertial and serious matter of classical physics responds to the laws of mechanics and the cosmos. There, we can also draw an equivalence between mass and information and we may add compressed information, enabling the order of arrangements, with the unfolding of cosmic time. Quantum gravity then becomes a questioning of the possibility of articulating a temporal quantum information and information helping cosmic time to emerge with a submerged order (that of compression).

11

Mesoscopic and Macroscopic

11.1. Between infrascopic and macroscopic quantum, the mesoscopic order

Modern science has produced two great narratives: the Big Bang and evolution. However, nothing is definitive and a new understanding is blossoming, derived from *the universal conjecture of emergence and communications*. The new approach embraces a conception of nature and the cosmos as instances in which things get organized and emerge because materially embodied components communicate by exchanging signals and at the elementary level of *quantum information*. Nevertheless, the articulation between quantum and the macroscopic level is not solved yet, neither in physics nor in biology.

In an article published in 2000, Robert Laughlin and his colleagues suggested [LAU 00] exploring an intermediate world between the level of quantum elementary processes and our empirical world, defined as classical in the physical sense. It has also been referred to as an ordinary, or rather, macroscopic world, in order to convey its large-scale magnitude. Now, is there an intermediate world between the microscopic (infrascopic) and the macroscopic? Laughlin, a physicist crowned by the Nobel Committee for his work on the quantum Hall effect, replied that yes, indeed there is. He actually described an intermediate scale of reality, the mesoscopic level, with a special remark: while it is true that the elementary level (particle or atom) has been thoroughly described by mathematical formulas, when the number of atoms increases, two things happen. First, an epistemological transition, due to the impossibility of making calculations in reasonable time for describing N-particle sets. Then, and this is the most interesting observation, a phenomenological transition translated by emerging phenomena (studied in chapters 7 and 8), such as the Hall effect, ferromagnetism, superconductivity, crystallization, etc. [LAU 00]. I suggest using the neologism "infrascopic" to designate the most elementary level of the quantum way. Now, the meaning of infrascopic is clear, because it refers to an order of reality in which the notion of space is no longer relevant.

The mesoscopic scale is a boundary for studying things at an intermediate level between the elementary particles (information) of the quantum (infrascopic) world, and our macroscopic scale. Emergences also occur at the mesoscopic scale, both in the world of inert materials and in the living world, with complex molecular "machineries" operating in a coordinated manner in cells, and capable of accomplishing various tasks: assembly of macromolecules, replication, management of genetic and epigenetic information, unexpected chemical reactions thanks to enzymes, nanomotors, cellular communication and membrane formation with two layers of lipids. Cell biology is a *mesoscopic biology*. As is also mesoscopic physics, which studies phenomena whose magnitude order lies between a few molecules and a micrometer. This is also the interval where the processes of a living cell take place, including the world of viruses. Molecular mechanisms occur at the nanometer scale in cells whose scale is the micrometer. Mesoscopic physics and biology have remained largely unexplored and, according to Laughlin, they are considered as still emerging disciplines.

The *mesoscopic sciences* inquire about the organizing principles operating at this intermediate scale; they are complexity sciences. *Mesoscopic structures* are based on organizational principles. The components of the mesoscopic level are ordered because they communicate and they exchange information. Living macromolecules adopt a special conformation which enables them to exercise their "molecular function" in an articulated way with the whole of the cellular machinery. This organization has remarkable structural stability as well as mechanisms capable of correcting disturbances. If we notice that this organization intervenes at ordinary temperatures, whereas in condensed matter, "highly ordered" properties require extremely low temperatures, then life is a sort of miracle. This is the example of Fermi liquids or superconductivity. At common temperature, a relative order is observed in the liquid and solid phases, as well as in liquid crystals, for instance.

One of the theories used in statistical physics is based on "energy basins", which are supposed to describe how components reach a state of equilibrium and maintain it. We can picture the situation as a billiard cloth whose surface is not flat, but contains tens or even hundreds of hollows. A ball is thrown, it keeps moving until it ends up in one of the hollows, where it finally becomes stable. This energetic landscape image makes it possible to "capture" the collective phenomena observed at mesoscopic scales. It is as if the components behaved competitively or hesitated before ending up occupying one state of the energy landscape, rather than another. Finally, the basins of the energetic landscape indicate how the distant parts of a system communicate (Laughlin), what "matches" the physics of communicative emergences.

The mesoscopic level is necessary to configure two large unresolved conjectures. First of all, the enigma of the emergence of the living, which is conceived as a transitional phase between an inert state and a living state of matter. The second question concerns the connection between the physical processes (inside the neurons), and the functioning of consciousness. A mesophysical approach is possible and was used by Roger Penrose, linking mental processes to mechanisms whose supports are microtubules, protein structures made of tubulin, thus belonging to the mesoscopic scale.

The mesoscopic level connects the elementary quantum and atomic world to a collective world capable of producing *mesoscopic emergences* which, when combined, produce *macroscopic emergences*, those in which we evolve. Let us calculate the number of elements for moving from the atom to a mesoscopic scale having the dimension of a living cell, that is to say, the micrometer. Three cubic centimeters of water contain 10^{23} water molecules. An easy calculation makes it possible to deduce that three cubic microns of water contain 10^{11} water molecules. That is to say, 100 billion. Therefore, macroscopic life is composed of cells whose number can reach a few hundred billion.

11.2. Horizontal and vertical emergences

Among emergences, some are *horizontal* and others *vertical.* At the mesoscopic scale, Fermi liquids or the conductivity of metals constitute horizontal emergences. There is no hierarchy in the physical order of communications. Often, it is enough to associate horizontal mesoscopic emergences in very large numbers to move towards macroscopic emergences. A wire micrometer conducts electricity exactly in the same way as a meter of this same wire. Contrary to this, the living cell represents a remarkable case of vertical mesoscopic emergence. This emergence is at the origin of life. One of the essential characteristics of the living is revealed by the extreme formal differentiation of structural components and of communication molecular signals. Viruses, bacteria and unicellular beings belong to the mesoscopic world. We cannot see them, except with electronic microscopes or by means of the pathologies they cause. In order to move on to the macroscopic scale, it is necessary to consider a second vertical emergence, the one going from the cell to the organism with its physiological functions, as well as its observable behavior in the environment. A differentiation takes place in this case, too. There are two to three hundred types of cells in the human organism.

The communicational emergence states that: *every emergence is linked to the selection of a communication mode performed by the components which exchange specific signals, be it in condensed matter, the living matter, organisms or even human societies.*

A second postulate completes the previous one: *a vertical emergence is based on hierarchy in the information processing system. That is to say, a natural operator is self-constituting: it organizes information, combines it, interprets it, transforms it, (transmutes it); it is what we call a semantic operator. It works in concert with an active (technical) operator, which sends control signals (as a reaction to the interpretation of received signals, if necessary) to the active peripheral components.*

In vertical systems, emerging purposes may emerge. These aims are determined by the transformations and interpretations of signals. Three devices must be differentiated and articulated: (1) reception, or rather, the capture of signals, (2) signal interpretation, and finally, (3) the response to all the interpreted signals. The interpretation depends on semantic operators and the answer (reaction) depends on technical operators. A sense (direction) in the action is chosen. The system produces meaning, this notion understood in the sense of “direction”, “purpose”, “orientation”, but also “sense”, when the system has reached the faculty of consciousness. The emergence of goals can be easily conceived for macroscopic systems such as societies and even referring to living systems, although this point is controversial. Once functions appear, purposes can be envisioned.

Purposes are also conceived at the mesoscopic level. The study of the processes located at macromolecular interfaces makes it possible to assign “mesoscopic functionalities” to these structures, be it in terms of metabolism, of construction of intracellular structures, of information transfer or transformation. Let us just illustrate by a few examples: the enzymes involved in the cycle of Krebs, or the functionality of hemoglobin for carrying oxygen, the microtubules in the organization of the cytoskeleton and RNA messengers for managing epigenetic information. Semantic processes are present at the cellular level and obviously in animals. The semiotic level emerges in man beyond the merely semantic stage. For example, the perception of a tree with its fruits is interpreted as a source of food. This is the semantic level. An animal will react instinctively. If the animal is hungry and the fruits suit him, it will eat them. Otherwise, it will turn away from the fruits. Man reacts by means of representation and deliberative reason. This is translated by the production of multiple purposes. For example, we could pick up some fruits for food, bring them home and share them, or we might as well collect some pounds in order to prepare fruit jam. *Kronos* is multiplied when it gets involved in human consciousness. (Cognitive) *Telos* then chooses one purpose by following a suppressive process. When a person chooses an option, it is by removing all other possibilities.

11.3. The signals associated with emergences

The philosophy of emergences studies all systems made of material components, which can be coordinated in order to produce physical properties or specific functions accessible to direct or indirect observation thanks to modern science technologies that can be used for exploring mesoscopic scales. Emergences rely on the communication of hardware components. A system that achieves emergences produces a field of communications, which enables components to exchange signals and to inform themselves (see Chapter 7). In complex systems involving hierarchy, a composite can become a communication signal involved in the genesis of an "informative field". This is the case of not only molecular mediators in cells (cAMP, hormones, etc.) but also viruses, complex composite systems and information vectors for the higher level of cells and organisms. Finally, in what concerns components, some become specialized for performing a transmission function in the system. This is the case of neurons, cells provided with fibers which enable them to transmit signals, or – at a lower level – scaffolding proteins that assemble, not only to build structures, but also to convey information at the core of the living cell with an involvement in transduction processes connecting the signals of mediators attached to membranes; in order to achieve functionalities or even to influence the epigenetic regulation inside the cell. Here are some points providing an overview on this issue, while completing the considerations developed in Chapter 7.

11.3.1. *Infrascopic fundamental matter and quantum interactions*

Quantum physics describes elementary communications of matter with our macroscopic world. However, it also describes communications between the material components of a system. The three types of fundamental interactions each have their signals for achieving communications, between the electronic interfaces and the photons, and between hadrons and electrons with the weak bosons. In short, hypothetical gluons guarantee "communications" between the quarks we cannot observe. Quantum fields are theoretical constructions that articulate communicating bosons and the gauges regulating communications. According to my hypotheses, these particle signals result from a selection process carried out by two other processes, affirmation and suppression, with the particles of our material world as final achievement, together with the gauges that adjust their settings. Matter is affirmation, anti-matter is suppression. Fundamental physical time results from the resolved dialectics between affirmation and suppression.

11.3.2. *Condensed matter and mesophysical emergences*

In a large number of emergences observed on inert materials, electrons play a decisive role. It is them who become the quanta in real communications fields, as may be a superfluid material, a simple conductor or the surface of a metal whose electrons behave following the rules of Fermi liquids. In these phenomena, the peripheral electrons adopt collective behaviors. In superconductivity, electrons associate in pairs and given the fact that the total spin is equivalent to one, these pairs behave like bosons and communicate more easily. In Fermi liquids, electrons combine so as to form quasi-particles by adopting a collective behavior. Phonons are also quasi-particles, as plasmons are. All these quasi-particles must be interpreted as composite communicating structures.

Dissipative structures and crystals are also mesoscopic and macroscopic horizontal emergences. The communications between components can thus be partly clarified. These are the resonances in dissipative structures; peripheral electrons in the ordered phases of matter (see Chapters 7 and 10).

11.3.3. *Molecular matter*

Chemistry relies on stable associations between atoms involving more than one electron. The process is very different from what happens with electrons in an electrical conductor. Another element is involved, not only in chemistry communications, but also in the mesoscopic emergences produced in the molecules of the living. This element is the proton, which is changeable in aqueous solutions and which is also involved in many hydrogen bonds, for example, in the bonds between the bases of DNA. G and C communicate with three hydrogen bonds, and there are two for the A and T pair. Thousands of molecular associations employ hydrogen bonds.

11.3.4. *Life at the mesoscopic level*

Horizontal emergences feature certain homogeneity in their components, as well as their communicating elements. Contrary to this, life is based on an extreme differentiation between structures and communicating elements. This reality was intuitively understood by Schrödinger, famous for inventing the oxymoronic notion of the aperiodic crystal as a support for heredity. The crystal refers to a molecular, aperiodic order, as well as to the complexity and inhomogeneous nature of living structures. Life emerged as a vertical mesoscopic emergence, with a hierarchical processing of information and the establishment of effective signals, the result of "semantic selection" enabling the received signals to produce meaning for the system and to elicit a coordinated component reaction.

The first stage of life was mesoscopic, with the appearance of the first cells capable of exerting a double function at the mesoscopic level. On the one hand, a metabolic and chemical regulation with the capture of luminous energy and the use of molecules derived from carbosynthesis, such as glucose. On the other hand, a regulation of information circulating in the medium which can be used, rejected or transformed. Viruses regulate genetic information in ecosystem environments. Bacteriophages circulate throughout the world of bacteria and contribute to the regulation of their genetic heritage. In response, bacteria develop strategies to "counterbalance" information, which may disrupt their information system. Some giant viruses can also regulate their genetic heritage against viriophage invasions.

In the mesoscopic living world, elements of the inert world such as protons and electrons grant communications, but some other, more complex signals have been chosen to take part in the emergence of life and provide it with a stable structural base in order to evolve with formal coherence within the context of informal chaos. Biochemical analyses have established a list of ubiquitous molecules that are found in distant species and which are involved in signal transmission by interacting within transduction systems. The best known of these systems employs the cyclic AMP as a mediator molecule, which activates a type of macromolecule capable of subsequently modifying other structures in turn. These are the protein kinases of the PKA family. Other kinases work with cyclic GMP as a mediator. Another transduction pathway uses ubiquitous signals of inositol derivatives: IP3, inositol triphosphate, PIP2, phosphatidyl inositol biphosphate and DAG, diacyl glycerol. These molecules work in concert with calcium ions, which are also involved in transduction via the membrane channels, opening up so as to let them go through the cell and transmit information, sometimes moving like a wave, which is a distinctive attribute in the communication fields contributing to the creation of order and emergence.

The study of information transformations, from DNA to proteins, through transcriptions and epigenetic regulations, suggests a selection of mechanisms and communication signals at the mesoscopic level. This is what enables the emergence of functionalities ensured by molecular structures. These mesoscopic functionalities cannot be assimilated to the macroscopic functions that physiology and ethology study. The fact that life spreads over some two or three billion years reveals the presence of a finality whose guiding principle is shared by cellular structures and the environment which is also structured like a field of communicating elements, in its own way. The *Telos* of molecular life is similar to the Spinozian conatus. In short, the mesoscopic life of a set of coordinated cells enables the emergence of macroscopic life. Cells have perfected their communication systems, as is shown by the results on the transduction systems, some of which are already present at the cellular level. This is the example of the cAMP-dependent pathway in *Saccharomyces cerevisiae* yeast, which enables these unicellular organisms to exchange signals for coordinating their growth.

11.3.5. *Macrobiology and macroscopic emergences*

Macrobiology studies living systems as they are revealed to us, at our own scale. We gain a direct understanding of macroscopic life and its functions, without going through the analysis and decomposition of tissues or cells. Our existence can continue, even if we ignore not only the quantum infrascopic world but also the mesoscopic living world. There truly is no need to know what our DNA, our RNA, proteins, molecular signals, our neurons and our myocytes do in order to reasonably lead our existence, work, walk, look, speak, say, do, think, ponder, reason, calculate, love, contemplate, eat, seek, build and invent. On the other hand, if we had to consciously manage all the information circulating in our cells, our psyche would be overwhelmed by a colossal flow of signals and elementary tasks to be solved. Although our senses cannot access mesoscopic things, our psyche can affect mesoscopic cells.

At the level of organisms, the world is macroscopic and the emergences that develop there, be they animals or plants, are based on a communication system capable of coordinating up to hundreds of billions of cells so that they can react in concert and cooperate in the achievement of macroscopic functions. Specific signals and adjusted structures enable this orchestration. At the molecular level, hormones transmit signals through the bloodstream; the best known are adrenaline or acetylcholine, which regulate the heart rate, blood pressure and the contraction of smooth muscles, among other things, or dopamine and serotonin, which are involved in the brain processes and regulate mood and tone, among other things. More complex hormones are also employed. These are small protein fragments, peptides, made up of a few amino acids. We know dozens of them, among which we can mention vasopressin, insulin, oxytocin and the famous endorphins, which are supposed to produce the feeling of pleasure in the brain. All these hormones act at a specific level, with a macroscopic effect, because the target of these signals is the organ, the macroscopic structure. The mediators of the nervous system operate at the level of synapses placed at the end of the nervous endings, while blood hormones become diffuse at the level of the capillaries in order to get attached to the target cells of the organ that is supposed to receive the information. Other hormones such as steroids have a different mechanism that regulates the body by operating at the level of the nucleus and of genetic processes.

Macroscopic life is the domain where functions with matching emerging purposes are accomplished. The decisive element for bringing out macroscopic life is the nervous system, which has been present since the appearance of the first animals and has enabled both the realization of semantic processes and cell coordination. Macroscopic life emerges by assembling fragments of mesoscopic life, provided that the enormous flow of information is coordinated (thanks to suppressive mechanisms, in particular) so as to provide coherent forms which enable living systems to take a hold on the environment and to capture meaningful signals. Life is not only a question of survival and success in the actions carried out in the environment. It is also a question of communication and interpretation of the signals with unlimited formal inventions. This involves fine-tuning and high adaptation in the information reception systems placed at the sensory interfaces, as well as a sharp perfecting of internal semantization and action determination systems, in view of accomplishing the purposes of specimens and species: esthetic forms, pheromones, flowers, natural scents, plumage, sounds, songs and language. The essence of life is to find a way by playing on Communication.

11.3.6. *Human societies and communicating forms*

If emergence is a universal concept, it is in the field of human ensembles that it is the most accessible, because the individual is ideally placed for observing emerging phenomena and for understanding himself as an actor, spectator and genealogical emergent subject in the emerging society. The consciousness that humanity is a succession of historical emergences is very recent, and only goes back two or three centuries, when modern philosophers chose History as an object of study, a History rich enough to be understood as a succession of epochs. Hegel, then Nietzsche and Heidegger lucidly questioned the phenomena of historical emergence. The first with dialectic and cultural-spiritual emergences; the second with his writings on the birth of tragedy and genealogy, among other things; the third with art as an emergent and meaningful form adopted by people becoming aware of the singularity of an era and the task at hand. As a general rule, the language and cultural forms that marked an era and spread in a civilization were the result of a process of collective selection carried out within societies by the actors and receivers of esthetic and intellectual forms. Communications reveal how societies work, what their temperaments are like, their values, the lifestyles that emerge and the aims that are pursued.

Through works of art and literature, people are able to talk to each other, communicate and, for us, who are thousands of years old, it is History and the people of an era who speak to us through these works. The information conveyed in a society can be of several types. Some pieces of information are directed towards the subjects and, once received and semantized, participate in the formation of the contents, knowledge, existential orientations and the sense of the intimate. The others are directed to the world. Information is then communicated to things by means of techniques and industries; they also serve as signals for commanding men, at the same time that signs are used for conveying meaning and culture.

11.3.7. *Ordering and interpreting systems in human emergences*

After the Neolithic, some societies were organized as big human enterprises thanks to a chosen language, because this was effective for achieving intentions. At the same time, language functions as a system for generating meaning, thought and knowledge systems helping people to orient themselves, to place themselves not in a physical, but a cultural and historical place-time, as well as in a personal intimacy. In the Middle Ages, command structures were located in castles and thought structures in universities as well as in monasteries. Thought and meaning were transmitted by theology at the time. In the modern era, the heroes of History and the statesmen transmit orders and meaning towards the society which they push in the direction they order. Artists, philosophers and scientists collect and organize information about nature, man and society. When Western Modernity emerged and was accomplished, the political order progressively diverged from the philosophical and religious order.

"High culture" texts contain three levels of understanding: (1) literal, (2) semantic and (3) symbolic. Humanity is the only species that can evolve in two worlds, one of material and sensory realities, and one of mental and spiritual constructions. The fourth level is that of mystery. It does not directly concern the philosophy of emergence because it refers to a non-material world, as well as singular verticality involving a very personal experience. In his *Tractatus*, Wittgentein concluded that we must remain silent about that which cannot be spoken about. What transcends language can only be expressed through a music that elevates thought and transfigures it. This was indeed the effect sought by Scriabin's latest research, including the quest for the *Mysterium* delivered through a few sketches and three movements, *Universe, Mankind* and *Transfiguration*, which were reworked by Nemtin.

A governor of an ancient city wants to build a temple. He pictures the image of the building. This is the literal level of representation at work. Then he thinks about the construction of the building and summons an architect who draws up a construction plan including the organization of the workforce and the orders given to the workers. The semantic level is used. Finally, the last level is semiotic. It concerns meaning and purposes. This temple will be placed at a specific place of the city, with the intention of being clearly appreciated. Then it will be ornamented with figures that may represent a deity or even a king. Some names or formulas will be inscribed on the pediment of the building. Going into the building, we will undoubtedly perceive a sacred space with statues, for instance, or with a geometric figure. The semiotic level inscribes the actions in an extended duration, with the concern about purpose, as well as a verticality, in the measure that the temple favors a communion with the higher realities represented by the figures and the symbols; thus emerges a society that cannot be reduced to a community of workers, administrators and entrepreneurs. Men are united by semantic and material links, but also by relationships endowed with meaning and symbols, and whose mediation requires specific forms of communication.

11.4. Biosemiotics, emergences and evolution

The paradigm of information represents a new direction in contemporary knowledge. Many researchers in theoretical biology have oriented themselves towards a conception of life in which information and its interpretation play a key role. A new field has recently emerged, "Biosemiotics". It has been developed by a small community of theorists in biology and neuroscience. Its aim is to show that life develops through fields of communication and processes that enable life to find meaning, a sense of direction. These processes are defined as semiotic, including the notion of "semiogenesis". An article took stock of this concept by developing eight theses at the time that it stressed the idea of a threshold in "semiogenesis" [KUL 09]; this threshold can be linked to an emergence: in the case of evolution, it involves a modification of the environment and the emergence of new species with their behaviors and modes of communication.

The first of the eight theses is central. It states that the separation between semiotics and non-semiotics covers the distinction between the living and the non-living. This statement is justified by the fact that the concepts of function and semiotic processes (operating on signs) are intertwined. In other words, every function which is present in a system depends on a process of signal processing. This semiotic process can be accompanied by an (internal or external) representation of the final state achieved after the implementation of the function. There is still an uncertainty concerning the determination of the threshold or the boundary that makes it possible to assign the existence of a semiotic process to a natural thing, and the *Telos* which accompanies it. According to Kull and his colleagues, in the absence of fundamental data on the processing and interpretation of signs (and signals), biology is an incomplete science; and this is the second thesis of this theoretical biology presentation.

The origins of life are based on semantic selection. Its traces can be found in information control systems located in bacteria and giant viruses. When macroscopic life emerged, functions emerged as well, resulting from an organized combination of mesoscopic features, which differentiated themselves when functions emerged. Hemoglobin ensures the oxygen transport functionality thanks to red blood cells, which are needed for the breathing function. Myofibrils are protein-like constructions endowed with a contractile function produced by the sarcomeres they are made of. Myocytes are the contractile cells of the muscle. They grant the production of muscle strength (which is a macroscopic function), whereas the myofibrils are mesoscopic. The achievement of the function is based on a close coordination of the millions of myocytes that constitute the muscle, with the sarcomeres at the mesoscopic scale, whose nano-mechanisms are coded in billions and must be synchronically coordinated. In the field of reproduction, the fertilization of an egg by sperm is a mesoscopic feature whose achievement requires the meeting of two sexual partners. This encounter is conceived as a macroscopic event that ethology and zoology study.

Capturing information is an essential macroscopic function. The achievement of these functions is accomplished thanks to the juxtaposition of structures and mesoscopic phenomena, in other words, the juxtaposition of functionalities. This is the case of retinal sticks, for example, or of the olfactory receptors located at the interfaces that represent the mucous membranes. Neurons make "mesoscopic information" circulate, thanks to their fibers and innumerable synapses, their mediators, their synaptic junctions, ion channels and electrical signal transmissions. The information supported by the mesoscopic functionalities (synapses, microtubules, electrical signals, etc.) combine and "add up" so as to bring out the higher functions of the nervous system. This manages the macroscopic information captured from the environment, or returned from the internal physiological milieu, in order to be interpreted by the individual. This is the case of emotions such as pleasure, fear, nervousness, or the feelings of well-being, and, on occasions, even fatigue.

As suggested by the fundamental principle of "biosemiotics", when we detect emerging mesoscopic features or emerging macroscopic functions, a purpose is revealed, which can be studied or assumed. Living systems are made up of elements that are organized according to the principle of vertical emergence. In an organ, the cells (mesoscopic elements) perform functions by juxtaposing and coordinating functionalities (for example, the simultaneous contraction of sarcomeres in order to produce muscular force). This coordination is more precise than in the most advanced artificial systems built by man. It is essential for generating macroscopic functions and relies on the transmission of signals, by means of command and a controlling device. This is the principle of vertical emergence.

Darwinian evolution needs to be rethought. The distinction between mesoscopic and macroscopic lives opens up a question field about evolution. In the animal world, life takes place at the macroscopic level in an ecosystem where sexual partners meet. Life is a game using the information that is emitted and captured, and the actions performed at the macroscopic level, a game that has lasted hundreds of millions of years. Nevertheless, life depends on mesoscopic rules (DNA, molecular processes) to perpetuate itself through the fusion of gametes. *Life uses power in the macroscopic world. Another power escapes from it, that of the mesoscopic and infrascopic world, with its elementary communications, functionalities and quantum entanglements*.

The mutation and the recombination supposed to explain the diversity of the living and even speciation are macroscopic notions unfit for the complete description of the fundamental mesoscopic processes, and in which quantum entanglement plays a decisive role. This is one of the deficits of the neo-Darwinian theory based on a rough understanding of matter that does not take the results of quantum physics into account. The other deficit of evolution comes from the preeminent place accorded to natural selection. If historicism is the Achilles heel of philosophy, as Heidegger and Eliade pointed out, then "natural selection" plays a symmetrical role in biology, by suggesting that most of the functional properties of life are the result of natural selection; rather similarly to History, which would be the magic key to account for human transformations.

Evolution is still a narrative of how species appear as emergences in a heterogeneous environment with geophysical conditions, plants, and molecular mesoscopic structures in which macroscopic specimens communicate and become involved in the game of life. The explanation for evolution is incomplete. Natural desire, generated by the reception of information, is presented as an engine for evolution. This was glimpsed by Bergson with his notion of vital impetus. Natural desire is an efficient cause whose action is guided by semantic processes. It intervenes in sexual selection, which cannot be limited to a competition of the males for the females. The animal world is the result of an implemented universal desire and works in concert with a living matter that begins to invent limitless forms. Desire accompanied by "biosemantic processes" enables a type of form selection and models species by fixing them in final forms which, as indicated by a new interpretation of sexual reproduction, does not seek the mixing of genomes at the service of diversity as fundamental result but, on the contrary, a convergence of species in stabilized morphologies. The species competition does the rest. Finally, we should remember that the mesoscopic, molecular, semantic and mechanical processes are often disconnected from life's contest, and take place independently of the macroscopic conditions at the heart of the organized system.

One last remark, though. The study of mesoscopic functionalities related to the organization of genetic and epigenetic regulations often uses a specific vocabulary for the macroscopic level. In addition, sometimes, mistakes or misinterpretations occur. Thus, the study of expression and interference systems in the regulation of RNAs and of viruses leads to the use of common terms, which are not necessarily appropriate for denoting the exact meaning of the processes discovered by overpowering genetic technologies.

We notice the use of a "warrior" vocabulary in the description of immune management processes deployed by bacteria against viruses, through the use of epigenetic methods aimed at preventing viral invasions, including CRISPR systems, one of which has been recently employed in genetic engineering due to its formidable effectiveness [LED 17]. In this description, it is a question of invasion, immune attack, defense and cell suicide. These few notions show a semantic continuity between the world of macroscopic functions and that of mesoscopic processes. Perhaps it would be necessary to consider some other continuity, either semantic or epistemological, between the mesoscopic living molecular world and quantum descriptions. In other words, it may not be legitimate to assume that viruses and bacteria have intentions which are specific to the macroscopic world of species. These remarks should be solely understood as programmatic markers for possible unpublished research in these fields.

A formula summarizes the epistemological positioning of biosemiotics, as well as that of emergence philosophy. *If physicalism attempts to naturalize semantic processes, we will then adopt the reverse position of semantizing natural processes.* This might represent a revolution in the field of scientific philosophy.

11.5. Physicalism and mental processes

What is the origin of mental processes, what is their physical essence? No one has a definitive answer but scientists can be divided according to two dominant conceptions: the physicalist monists and the dualists. Physicalism follows the principle of considering that, in the last resort, everything can be reduced to explanations and physical descriptions. Physicalism is often considered a type of materialism. Nevertheless, the two doctrines have a different story. Modern materialism was born during the 18th Century and was essentially conceived as a sensualism. It was a philosophical doctrine that opposed Cartesian ontological dualism, with its two substances, the extended body and thought. Physicalism was conceived differently because it relied upon a "scientific ontology", that is to say, upon the conception of matter in the way that this transpires from the results obtained by physicists, through experiments, calculations and formalizations. Physicalism represents the scholarly materialism of the 20th Century. Opposite to this, there is a school of thought introduced as a "non-physicalism" with several options. The debates on physicalism occupy hundreds of books with developments in multiple directions, and often great technicality, as well as a lot of nuances in every field. The best known physicalists are Daniel Dennett and Jerry Fodor. Confronted against them, Thomas Nagel and David Chalmers are critical of physicalism, without nonetheless being convinced dualists. Eccles and Popper were dualists, with a differently appreciated theory, the interactionist dualism based on an extrapolation from quantum duality. It is impossible to present all the discussions and conceptions on this subject considering the hundreds of works published. I am just suggesting some ways to understand how the mind emerges from the brain.

Mental contents are produced by brain functions that originate at the mesoscopic level, in the same way that mechanical force is the product of the muscular function. This function is conceived as a macroscopic emergence, due to the coordination of mesoscopic processes granted by each of the activated myocytes, whose molecular mechanisms are produced in the sarcomeres when these are synchronized. Mental contents can be expressed through the body. Language is produced by the pulmonary breath, which makes the vocal cords vibrate. Sound is then shaped according to the position of the lips, language and the movements producing the transcribed phonemes, which follow the intention of transmitting mental content. These phonemes are sometimes associated with letters and symbols combined from different known alphabets. Language is a content of thought that does not need to be materialized orally in order to express itself. We can talk to ourselves silently. The mental contents are caused by the coordination of mesoscopic processes carried out at the neuronal level, thanks to fibers and synapses. Self-awareness emerges when a person manages to communicate with themselves, that is to say, when they express a mental content and are able to receive it by reflexivity, mentally read it and interpret it.

Nerve cells add up their effects to achieve macroscopic and basic mental functions, which can express words, signs and perceptions, such as the sensation of color or bitterness. Vertical emergence leads to semantic and semiotic structures that produce sense and meaning. Why are these mental forms possible? As information is already present in matter at the quantum level and at the mesoscopic level, forms emerge as combinations of information, vibrations and coordinated resonances. These neuronal mesoscopic processes add up in order to build mental contents. Quantum matter is a set of communicating interfaces. Among these interfaces we can mention the superposition of basic states and entanglements. A "content and information" duality is therefore pre-existing at the very level of "quantum matter". Then, the neuronal world is in correspondence with the fundamental duality in "fundamental quantum matter".

These considerations point to the impasses in the sciences of the mind which, claiming materialism or not, are limited by an incomplete knowledge of the fundamental processes of matter or by a simplified vision of matter inherited from atomism and classical thermodynamics. To put it simply, the physicalist does not know the scope and importance of information in matter with its communications. The dualist goes astray in a non-physical conception of information by sometimes inventing hypothetically immaterial information. The mind actually emerges from matter because information is in matter and that matter has evolved throughout subsequent emergences until reaching the human brain. Thought emerged as an active macroscopic information processing system built with information itself as leverage and raw matter which was transmuted, since it constituted a form of energy. It is for this reason that a computer cannot acquire consciousness. For that to be possible, we would have to conceive software and a hard disk memory transmuting themselves to conform the processor and the motherboard. The theories of transhumanism and singularity are in an ontological impasse. The world of computers communicates with signals, but digital networks, although containing operational centers and singularities, cannot produce an ontological vertical emergence as cells and functional organisms endowed with mental capacities can do. Technological singularity can only lead to an artificial pseudo-emergence that is completely out of proportion compared with what is achieved by nature.

To summarize this question, it is stated that neuronal mesoscopic systems are semantic units endowed with functionalities. The juxtaposition of these mesoscopic units accompanied by hierarchy leads to the macroscopic semantic systems whose functions produce mental contents.

11.6. Some considerations on emergent worlds

11.6.1. *Gravity, AdS/CFT, megascopic emergence cosmos*

The cosmos can be conceived as the result of an immense material process that has emerged. The cosmos encompasses us and must be described as megascopic. Our universe comprises four levels, the elementary and quantum microscopic (infrascopic), the mesoscopic in matter and life, the macroscopic in the functions of the living, the species, human societies, the Earth, and finally, the megascopic, designating the extended cosmos with arranged masses and the world scene, which is also emerging. It would be presumptuous to suggest here the elements for an alternative cosmology to that of Einstein. I will stick to a few suggestions, including the idea of a cosmos conceived as an emergence. In order to grasp the difference, relativistic cosmogony treats extension on the same ontophysical plane as matter. This makes it necessary to treat space–time with a physics of arrangements, while allowing for the description of a universe in expansion or contraction. The emergent cosmology that I propose is based on an ontophysical difference between extension and matter. This leads us to conceive two orders in the cosmos, one linked to the gravity-matter that arranges masses, and the other to a physics that regulates communications between the material elements. Then, according to the principle of horizontal emergence (which acknowledges the singularities of the cosmological information related to the material masses of the universe), these two orders are co-emergent. Two instances order the scene and its actors. These are gravity and magnetism.

In this context, the gauge/gravity duality or CFT/AdS makes it possible to design this co-emergence by establishing a correspondence between condensed information in matter (CFT) and a nomological description of the scope (AdS), with an additional dimension, the one we can interpret as a degree of freedom enabling masses separated by space-gravity to communicate. The gauge/gravity duality represents an alternative formulation of the principle of equivalence governing relativistic cosmology, the equivalence between matter and geometry, which means that matter is transformed into geometry and the other way around. Megascopic emergence is generated by moving directly from the infrascopic to the megascopic. It is the strangest of physical enigmas, the one that conjectures quantum cosmology and which we do not know how to solve. Quantum gravity is the answer to the riddle, but we do not know what gravity is. Our planets were formed with two processes, one of quantum essence (cooling), and the other of gravitational essence. In short, we must assume that the conjecture of quantum gravity is a mystery.

The portrayal of the megascopic order would not be complete without the scene of communications governed by the EM field and regulated by the speed c of light. In one second, the moon and the Earth are connected. In turn, the gravitational scene of arrangements is regulated by the constant G. Separated by billions of kilometers, the stars influence each other. This order is immeasurable with that of the human condition in the macroscopic world. Thus, humanity can transform the planet through technique because planet Earth is in the macroscopic order, but humanity can have no impact on the megascopic cosmos extended over billions of light years.

11.6.2. *The emergence of the sacred and Being*

Man's predilection for the sacred reveals the knowledge of a universal principle present in the cosmos (meaning order against chaos). The sacred is a hierophany endowing a place with a certain order and power. The sacred presents itself in different manners in societies and in History, with some unchanging characters: the opposition between order and chaos, communion and verticality, reference point in existence and supporting point for consciousness which, in time, is doomed to disturbance by the different noises associated with disorder in the world. The sacred means to privilege formal or symbolic efficacy and to keep it intact, in such a way that this efficacy can be communicated without alteration due to time or to the chaos of the extended world. The sacred is a necessary knot for order to emerge in society, and keeps them from chaos and disorder linked to the power of *Kronos*, the dividing evil, etc.

In the sacred, there can be no emergence of civilization. When societies neglect the sacred, they perish and liquefy in the formlessness of chaos or they must provoke the emergence of new arrangements capable of representing something sacred linked to new historical aspirations. In Modern Europe, the nation became sacred, but this sacredness did not last for long. A couple of centuries, to say the most. For the sacred to emerge, "time solidness" is necessary, although our era is going through the liquidity of activities and cultures, since it venerates change.

11.7. The order of time at the four scales of the universe

This last section announces a vast project seeking to characterize emergence processes as linked to the considerations about time that we developed in the two previous chapters. Observations have shown that there are different time scales, as there are different size scales. These times record events within a certain lapse, but they also reveal the cyclic processes that we may conceive as the expressions of a small-scale *kosmos*; in the world of quantum processes, that of the phases of matter, of the living cells, of animals, of civilizations. The four orders, infrascopic, mesoscopic, macroscopic and megascopic work with length scales and periods featuring the cycles produced by the vibrations and other resonances capable of transferring information.

Let us not forget that there are two times in cosmos, one associated with gravity and arrangements, and the other with the communications scene, which separates the emission of a photon from its reception, and which signals the transformation of systems when it becomes biographical. The time cycles measure the revolutions of planets around the sun. However, it is another time that signals the exchanges and communications in our macroscopic world. Time cycles determine the majority of processes in matter and in life, at very different scales. The analysis of the connection between time and communication would deserve a complete essay.

For instance, the vibrations of molecular components at the bottom of the mesoscopic scale (some 10 nanometers) take place in short periods, milliseconds or microseconds, or even less. This is the case of molecular processes in sarcomeres, and the cyclical reactions at the cell level. The hormonal cycles belong to the macroscopic dimension, at our scale, with duration measured in hours. The menstrual female cycle lasts for a moon-cycle. There is a relation between the period of cycles and the orders of magnitudes, infra-, meso- and macro-scopic.

The connection between these periods and communication is often related to the order of time.

Finally, we cannot avoid the distinction between the cycles of arrangements and of communications. These two determinations are of a very different essence, but articulate with each other, which contributes to the formal and dynamic richness of our world. One book might not suffice to describe these processes at all the scales that science can experiment.

12

Epilogue on Forthcoming Science

12.1. Modernity is achieved

This journey into the universe of natural quantum communications and the order in matter opens new and unprecedented, even unexpected, perspectives for science and philosophy. A split is emerging. Knowledge is about to shift in the same way it did some four centuries ago, after Galileo and Newton. We are moving from the modern to the postmodern era, from the era of mechanisms to the era of communications and information.

Modern science was built upon the principle of the retraction of the "perceiving subject" (embodied by Aristotle), in favor of the experimental subject (Galileo and Newton). Nature was placed at "gnoseological distance" and became a suitable material for measurement and use. At the same time, the philosophical subject asserted itself as an actor on the stage, which led to the removal of God and the "death of God". Men then conceived existence as a problem to be surmounted and nature as an enigma to be solved, instead of conceiving the destination of man as a mystery to be lived and nature as a mystery to be explained. The history of knowledge cannot avoid the epic of the modern subject which asserted itself after the Renaissance, by manipulating nature and establishing a distance from God, and afterwards, in the late 19th Century, "divinized" society and history. This same subject was later questioned and had to face the pressure of social norms, as well as the information and noises which have steadily increased since 1945, and the advent of media and the digital. The 20th Century was crossed by suspicion and deconstruction (Lacan, Derrida, Deleuze, Lyotard, etc.).

The paradigm that has emerged in this book articulates time, emergences and communication. It concerns us closely because we are at the heart of a global experience. Never have communications been so intense: not only generating noise, but also being sources of new emergences. The year 2017 proved to be the end of an old world which had emerged in 1945. The emerging world of the last 70 years seems to be disintegrating, without us being able to bet on a new, unprecedented emergence, or the liquefaction of civilizations in a formless and informational chaos. History is uncertain and is played between *Kronos* and *Telos*. Recent political events in the United States, the United Kingdom and France are witness to this. Digital communications have favored the dissolution of the old world.

Nature, as explored by experimental technologies, suggests the return of *Telos* through physics and soon biology. It is not the new facts that change our perspective, but the interpretation of available theories. The understanding of natural emergences in matter is based on the interpretation of quantum physics, informational order and the dual orientation of time, with the dialectical game between *Kronos* and *Telos*. Life and evolution must also be understood as emergences. So let us now summarize the main principles of the "scientific philosophy" that we have outlined in this book.

First, we must distinguish between the field of *actions and arrangements*, and that of *communications* [DUG 17]. Then, we have to take into account the dimension of time and consider two system categories: those which have emerged and are endowed with a stable or even stationary order, and those which are emerging thanks to communications and the interaction of components with noise and information, which regulate the ensemble and make it possible for order to emerge (or not) between the two poles of time, *Kronos* and *Telos*. This results in the coordination of *actions and arrangements* as well as order in *communications*. Expressive and receptive structures receive signals that can be read, interpreted and transcoded, or even "transmuted", with nuclei capable of centralizing and managing compressed information and, if necessary, assigning it with a hierarchy (this is the case in human beings and societies).

The philosophy of emergence offers a new look on natural things. Matter is neither a shapeless substance nor a set of unruly atoms agitating like a clinamen, randomly achieving its structure. Matter, whether belonging to the cosmos or to our world with its multiple phases, must be considered as an emerging structure as well as "plastic matter" (in the sense of plasticity), capable of producing emergences. Matter communicates at its level, is regulated by the dual direction of time (energy), *Kronos* and *Telos*, informs itself and is ordered. It became ordered from the phases of matter until the development of life and man, a species capable of understanding how nature emerged and how it fits into a social and historical order, by creating civilizations as large-scale-emerging ensembles. Man is destined to communicate and to make consciousness emerge. The modern era is coming to an end. Hegel stated that the essence of man is to work, not without glimpsing a spiritual destination for the human. The opening era is under the aegis of a new formula: the essence of man is to act and communicate. However, the highest human destination is communion through consciousness. *If communication concerns decompressed information, then communion is conceived as a resonance between compressed pieces of information.*

The emergent (emerged) structures are deployed in the material extent at the time that they are connected to an order described as submerged. How not to evoke the unconscious as a submerged dynamic structure specific to man, and connected to the emergent or emerging consciousness? The unconscious is structured, ordered but also indocile; in other words, the unconscious brings together the two polarities of time which are submerged *Kronos* and *Telos*. By connecting the "id" of the unconscious to the libido, Freud glimpsed *Kronos* as plunged under consciousness. By conceiving the superego, the trace of *Telos* was detected. There is still a future for the philosophy of the unconscious.

12.2. Unfinished notes on the cosmos, physics and metaphysics

A new cosmology emerges, with an unprecedented philosophy of matter. Quantum physics tells us that the tiny details of nature cannot be described with the classical notions of particle, force, material point and trajectory. The most important notions of Newtonian classical mechanics are no longer valid in the quantum universe. The spatiotemporal extent: the mass, force and trajectories, are all characters of the world having emerged from "primordial quantum matter". Time is also a character having emerged from nature and the cosmos. Information is equivalent to energy and matter. Information is also the source of time.

The study of condensed matter with the tools of quantum physics makes it possible to formulate the hypothesis of an equivalence between matter and information, as established by Wen (see Chapter 6). The analysis of Schrödinger's equation leads to a split in representation, with a duality of form and energy; this duality possibly being a double representation (as can be the wave-duality corpuscle) or rather an equivalence. In the case of an equivalence between energy and information, it is necessary to study the possibility that energy is transformed into information, and vice versa, otherwise we might confront nomological duality. As can be deduced from the other canonical equation of the quantum world, the $E = h\nu$ formula, where h represents the quantum of information or "informaction" and ν represents the frequency which can be interpreted as a "formal determination" of information, because it is a whole number. Another equivalence or duality is explained by the $E = mc^2$ formula. In this equation, mass is equivalent to energy. This equivalence was verified in well-known nuclear phenomena, such as controlled chain reactions in civil nuclear power or the explosion of the H bomb. Finally, general relativity and its $R = T$ formula translate an equivalence between mass and geometry, which is the form of extension. In addition, geometry (be it supposed real or deduced as nomology) is a dynamic that connects us to the energy and a certain order specified by cosmological time, *Kosmos* linked to massive matter and arranged to form the universe.

In summary, time is equivalent to energy. In addition, energy is equivalent to information. In the quantum world, there is a *real* equivalence between matter and information. In the cosmos, there is an *ideal* equivalence between mass and geometry.

At the quantum scale, we see two faces of the real, two orientations of matter, drawn toward expression and reception, these two poles being articulated with the directions toward the future and the past. CPT symmetry rules the quantum world. The expressed image corresponds to the received image built, thanks to a mirror-like reversal (P) of energy time (T) and charge (C). P represents the form, T represents the dynamics and C represents a nomological constant. The changeable phenomenal nature is based on an excess of matter, which leads to an excess of time directed toward the future.

The primordial quantum matter is ordered according to one principle (gravity, magnetism, or both?) and the emerging macroscopic world. The double polarity discovered in the quantum world of information with positive-energy particles and negative-energy antiparticles results in a duality between expressed *Kronos* and manifested *Telos*. Natures are determined by submerged *Kronos* and submerged *Telos*. The world emerges with a complementarity between the two faces governed by *Kronos* (expression and action) and *Telos* (reception and purpose). This complementarity is translated by the characters, forms and functions emerging from one side, and the other side, by submerged nuclei regulating communications and coordinating operations.

The world scene is covered with a thick curtain whose drapes have been raised, and we see the universe from an unprecedented angle, without the surprise being complete. It was predictable that the veiled reality would not be destined to conceal its enigmas forever. Science has always advanced by pushing the horizon of enigmas, while meditating on the enigma of the horizon.

Quantum physics generates a shift that is also a jostle. It leads us to reflect upon nature in a different manner, no longer in terms of mechanics, force, trajectories and objects, but in terms of process, information and emergence involving communication fields, properties, functions and, on the inner side, contents, those of underlying, submerged order. Quantum physics enables modern science to break out of its lethargic sleep after four centuries of mechanistic and atomistic conceptions. However, we must be careful because quantum physics is also a powerful euphoric that can generate many illusions, if it is not surpassed by an all-embracing synthesis which includes global order.

12.3. Communicational emergences represented in a figure

Figure 12.1 summarizes the scientific philosophy of communicational emergence, whose main results were exposed in Chapters 6, 7 and 11. It makes it possible to visualize two aspects functioning together in submerged systems: from matter to man and societies, while also exploring cells and animals. We must bear an important observation in mind: for material systems, only lines (1) and (2) are used. Lines (3) and (4) concern living systems and human sciences.

(1) Communications field ←← Action field

chosen signals ↑

↑↓

(2) Expressive and receptive structures Active technical structures

↑ ↓

(3) Communicated sense – Interpreted sense ↑

↑ ↓

(4) Information-generative nuclei **Operative central nuclei**

{Submerged order}

Figure 12.1. *Summarizing diagram*

The left side of the figure shows the importance of the communications and the place of generating nuclei, which endow the signals with efficiency in the fields of transmission, reception as well as interpretation. These processes are complementary to the operations and actions which, in inert matter, are barely simple reactions, whereas in autonomous systems, strategies and purposes are followed and sorted following several rules. There is natural selection in the action and a semantic natural selection that operates on the signals, with receptions and expressions. In order to evolve, a living system has to face physical rules and mechanical competition so as to win in the field. However, it also has to deal with the rules of information, noise and disturbances and know how to drive strategies, by capturing signals and developing strategies related to biosemiotic processes.

Science knows how to analyze the signals, the molecules, but as a result of splitting the components, it lost sight of the organization of information nuclei, which are signal-generating structures as well as interpretive devices. Analytical science destroys most natural communications and is powerless to understand the totality and emergences, which are of a dual nature, simultaneously {mechanical/functional} and {communicative/informational}. Not forgetting the genesis of submerged nuclei, which, for the time being, remains inaccessible to knowledge. Submerged nature loves to hide; we can suggest paraphrasing Heraclitus' famous formula.

12.4. The metamorphosis of the subject

The scientific philosophy of emergences and communications makes it possible to reintroduce the subject as "substance underlying phenomena", not only in its anthropological version but also as a universal pole that we encounter at all of nature's scales, from infrascopic quantum matter to the megascopic cosmos, including the macroscopic world in which we live. The following figure shows this new manner of thinking about things from an extended objective pole, world that is sensitive and analyzed by means of phenomena, object studies, science and experience. The "subjective" pole is made from primordial quantum matter, ordered by two "temporal intentions", *Kronos* and *Telos*. To the left of the figure, excessive *Kronos* produces the forms expressed in the extended world. These forms are then communicated and received, because they rejoin direction and maybe also the attraction produced by *Telos*-governed matter.

Objective pole, expression and communication field

Expression → → → Reception

↓ ↑↓ ↑

telos ↑↓ *kronos*

Kronos Telos

Subjective pole

Figure 12.2. *Subjectivity and objectivity*

The reintroduction of the inner pole of subjectivity concerns matter, nature and cosmos, and not only man. This conception goes far beyond the objectivist principles of modern Newtonian science. The subjective pole is submerged, as opposed to the manifest emergent pole. The submerged face with the dialectics of *Kronos* and *Telos* is involved to the same extent as the emergent face, with natural selection governed by the biographical *Kronos*.

12.5. From *Kosmos* to *Logos*

In Greek Antiquity, at the dawn of philosophy, *Kosmos* was already known to the Gnostics, who believed in a universal mind ruling the world, the *Nous* from Anaxagoras. Among the Orientals, this was *Tao*, understood as the principle and universal nucleus, which determines order in nature. These two principles, *Nous* or *Tao*, are derived from the contemplative gnosis based on empathy with the cosmos. The contemplation of nature allows the feeling of a cosmic order to emerge in consciousness, an order based on the same principles as the spiritual order emerging in existence and reflected by the cosmos.

The human destination moves forward to the rhythm of *Kronos*, a principle that "upsets" the cosmological, natural, biological and human order. Can the *Telos* of human destiny be an emergence from incarnated *Kronos*? The path toward *supernatural* remains open, here under the "form" of the mystery of time. *Logos* in correspondence with *Kosmos* and the Spirit leads to an unavoidable question: could this *Logos* possibly be co-generated by *Kronos* and *Telos*? As the Spirit, co-begotten by the Father and the Son? This is the mystery of the human destiny and the encounter with the incarnate Spirit, as a revelation of the submerged Being (a kind of universal nucleus). Natural theology could be bonded with theology as revealed by the Ancients, but the ontology of time was also unveiled by these modern prophets, who were the philosophers of consciousness, from the 19th to the 20th Century.

Philosophy opens the way to the Spirit in biographical time because it is based on the experience of the subject or *Dasein*. A part of theology is based on the experience of the mystery, which is interpreted as an exit from time.

12.6. Epilogue on three ongoing scientific revolutions

1) *The quantum revolution of information.* Quantum physics is currently considered as a physics of information, by the most insightful scholars. What is new is the shift from information to communication. Quanta of information are shared, thanks to a double dynamic linking the material things when the processes located at the quantum interfaces take place. CPT symmetry describes these two-way dynamics, expression and reception. This new understanding of the topic leads us to a dual description with expressed information; the most important part remains submerged and positions itself as contents, as opposed to transmitted and phenomenal information. However, we have to remain cautious and not allow ourselves to be deluded by a kind of Quantum Grail which can explain everything, from life to consciousness. The branches of quantum physics can describe how matter communicates, but not how submerged information is ordered. The only practical clue is entanglement, a property that goes beyond the mere study of communications and that sparkles systemic interest.

2) *The semantic revolution of emergence.* We are now certain that inert or living matter communicates. But what about emergences, and especially what about consciousness emergence, a property about

which man is the only species who can say something? Consciousness is based on semantic contents that are mental processes. The physicalist program attempts to naturalize semantic processes. The project of communicational emergence that I have just introduced proposes the opposite approach, aiming at semantizing natural processes. This does not necessarily mean granting a rudiment of consciousness to matter, but it entails describing the specific fields of communication of material systems by analyzing their phases, chemical reactions and the molecules of the living and cells.

The semantic revolution involves three aspects. First, the systemic conjecture on the ensembles of components, whose observable functions can only be explained by coordinating these same components. The second part concerns evolution and an unprecedented principle of selection of the signals within a system, as well as the environment in which these systems intervene. Thus, nature evolves following a universal principle, that of semantic Darwinism. Nature's systems evolve by making signals evolve as well, so as to ultimately communicate with a kind of "semantic convergence". The signals respond to an efficacy in intentions, and the means deployed are maintained and selected from an ensemble in which noise also reigns and is itself composed of signals, most often disruptive, non-integrated ones, which can eventually become the source of innovation if they are inserted into a new semantic order.

The third aspect of this revolution is the thesis of "generating nuclei", which generate signals that have a capacity to convey meaning or information when they are received. Chapters 6 and 7 explained that matter communicates with the bosons generated by the gauges, that cells use a limited series of (metabolically generated) molecular mediators and finally that the men have designed effective languages showing invariances, according to the theory of generative grammar. The signals that are specific to an emergence level move and constitute a field that can be defined as semantic or pre-semantic.

3) *Cosmological and biographical revolution of time.* For conventional physics, the fundamental laws of the cosmos are reversible, whereas matter is governed by an arrow of time oriented toward the future. For a physicist of the 20th Century, irreversibility

must be deduced from reversible laws. That is to say, if *Kosmos* represents classical determinism and *Kronos* indeterminism associated with the irreversible, then in the conventional conception, *Kronos* must be understood and interpreted from *Kosmos*. Now, in my study, I introduced an additional temporality, *Telos*, whose orientation is reversed, while being asymmetrical as regards *Kronos*. This leads to a Copernican revolution of time, by proposing that the "objective Kosmos" of the gravitational cosmos emerges from a resolute dialectics between *Kronos* and *Telos*. This conception makes cosmological time the "character of an emergent order". If time is an emergence, then there can be no Big Bang because there is no origin for the universal time which was not present in the universe at the beginning. Then, time is what happens at the end of a universal process whose result is the cosmos. Nevertheless, it may well be that the question about the origin of the universe has no scientific or philosophical answer. If this is the case, this question might not be necessary.

The biographical and cosmological revolution is also present in the highest human sphere, that of the experience of *Dasein*, but also a mystery of time interpreted as a "triple game" between *Kronos*, *Telos* and *Kosmos*. It is premature to draw theological conjectures. Let us remember that *Telos* is related to *Logos* and the supra-human spirit, *Kronos* to power and *Kosmos* to the Being. *Kairos* insinuates itself into this transcendental game determined by the essence of time and which is revealed in the experience of *Dasein*.

Conclusion

Final act

Some four centuries ago, Newton and Leibniz invented differential calculus without necessarily sharing the same vision about nature. Scientific modernity was propelled by Newton with the principle of space measurement and calculation, time and forces. Modern physics was constituted by obscuring communicating nature while placing it as a thing that can be manipulated. Modern science reduced matter and nature to silence. The subject retracted. In the midst of a scientific and conquering 20th Century, Monod affirmed the loneliness of man who randomly emerged from the immensity of the cosmos. Truth be told, it was the modern Promethean man who placed himself in a situation of indifference, refusing to commune with nature while using it without limits. Now, this nature contains the principles that enabled the emergence of man who, while nature came to help amidst contingent circumstances, testified in favor of an enlightened chance, somehow guided by the evolutionary *Telos*. At the beginning of the 21st Century, we know that modernity is over and that we can once again listen to the things of nature, matter, plants and animals, perhaps with the old covenant at the end of the way, reencountered in its entire splendor, with its procession of revelations and communions.

> "In the world of composites, the world of matter, everything is full, which means that all matter is interlinked. If there were empty space, a body might move in it without affecting any other body; but that is not how things stand. In a plenum, any movement must have an effect on distant bodies: the greater the distance, the smaller the effect, but always some effect. Therefore, each body is affected by the bodies that touch it, and feels some effects of everything that happens to them; but also through them, it also feels the effects of all the bodies that touch them, and so on, so that such communication extends indefinitely. As a result, each body feels the effects of everything that happens in the universe, so that he who sees everything could read off from each body what is happening everywhere; and indeed, because he could see in its present state what is distant both in space and in time, he could read also what has happened and what will happen… *sumpnoia panta*, Hippocrates used to say. But a soul can read within itself only what is represented there distinctly; it could never bring out all at once everything that is folded into it, because its folds go on to infinity". (Leibniz, Monadology, section 61)

Prospect

A final diagram summarizes what connects our understanding of the world to the concepts of nature used by the Greeks and the Medievals (columns 1 and 2), but also what is specific to our understanding based on the results of modern science, interpreted by assigning a decisive role to communications and to the experience of nature (columns 3 and 4). Thus, nature has a veiled face, as the Ancients knew, but nature also communicates, perceives, "learns" and transforms itself as indicated by the evolution of the species and the emergence of complicated systems. For the Ancients, time is already there, whereas in the diagram that I propose, time is of Trinitarian nature and is outlined as an emergence. The Ancients understood themselves as children do, even with the toys of time, while Moderns are the actors of time with progress, as well as the Gnostics of time with the triplet creation, revelation and redemption. This diagram in Figure C.1 summarizes the new research path opened by this essay:

energeia	*existentia*	emergence	*Kronos*
(idea) entelechia	*essentia*	submergence	*Kosmos* (*Logos*)
Ousia	*Ente*	Quantum being	*Telos*

Figure C.1. *Correspondences between time and ontology*

Bibliography

[BAL 10] BALAZUT J., "L'homme, l'animal et la question du monde chez Heidegger", *Klesis – Revue philosophique*, vol. 16, pp. 8–26, 2010.

[BIS 03] BISHOP R.C., "Brussels-Austin nonequilibrium statistical mechanics in the later years: large Poincare systems and Rigged Hilbert Space", arXiv eprint 0304020, 2003.

[BIS 05] BISHOP R.C., "Resonances, unstable systems and irreversibility: matter meets mind", *International Journal of Theoretical Physics*, vol. 44, pp. 1879–1888, 2005.

[BIS 06] BISHOP R.C., ATMANSPACHER H., "Contextual emergence in the description of properties", *Foundations of Physics*, vol. 36, pp. 1753–1777, 2006.

[DEU 14] DEUTSCH D., MARLETTO C., "Constructor theory of information", *Proceedings of the Royal Society A*, vol. 471, 2014.

[DIR 32] DIRAC P.A.M., *Les principes de la mécanique quantique*, PUF, Paris, 1932.

[DUG 96] DUGUÉ B., Procès et miroirs, PhD thesis, University of Poitiers, 1996.

[DUG 98] DUGUÉ B., *L'Expressionnisme, prolégomènes à une métaphysique des temps nouveaux*, L'Harmattan, Paris, 1998.

[DUG 14] DUGUÉ B., *Le sacre du vivant*, Temps présent, Paris, 2014.

[DUG 17] DUGUÉ B., *Information and the World Stage*, ISTE Ltd, London and John Wiley & Sons, New York, 2017.

[DUR 10] DURT T., "Anthropomorphic quantum Darwinism as an explanation for classicality", *Foundations of Science*, vol. 15, no. 2, pp. 177–197, 2010.

[EVE 10] EVERETT D.L., *Le monde ignoré des Indiens pirahãs*, Flammarion, Paris, 2010.

[FEY 79] FEYNMAN R.P., *Mécanique quantique*, InterEditions, Paris, 1979.

[GIL 96] GILBERT S.F., OPITZ J.M., RAFF R.A., "Resynthetizing evolutionary and developmental biology", *Developmental Biology*, vol. 173, pp. 357–372, 1996.

[HAD 04] HADOT P., *Le voile d'Isis. Essai sur l'histoire de l'idée de nature*, Gallimard, Paris, 2004.

[HEI 67] HEIDEGGER M., *What is a Thing?*, Gateway, Washington, 1967.

[HEI 78] HEIDEGGER M., *Being and Time*, Wiley-Blackwell, Hoboken, 1978.

[KAS 14a] KASTNER R., "The Emergence of Spacetime: Transactions and Causal Sets", in LICATA I. (ed.), *Beyond Peaceful Coexistence: The Emergence of Space, Time and Quantum*, World Scientific, London, 2014.

[KAS 14b] KASTNER R., "Einselection' of pointer observables: the new H-theorem", *Studies in History and Philosophy of Modern Science, Part A*, vol. 48, pp. 56–58, 2014.

[KLE 96] KLEMPERER V., *LTI, la langue du III*[ème] *Reich*, Albin Michel, Paris, 1996.

[KUL 09] KULL K., DEACON T., EMMECHE C. *et al.*, "Theses on biosemiotics: prolegoma to a theoretical biology", *Biological Theory*, vol. 4, p. 167, 2009.

[LAU 00] LAUGHLIN R., PINES D., SCHMALIAN J. *et al.*, "The middle way", *Proceedings of the National Academy of Sciences*, vol. 97, pp. 32–37, 2000.

[LED 17] LEDFORD H., "CRISPR's mysteries", *Nature*, vol. 541, pp. 280–282, 2017.

[LEI 91] LEIBNIZ G.W., *Monadologie*, Livre de Poche, Paris, 1991.

[LLO 12] LLOYD S., "The Universe as Quantum Computer", in ZENIL H. (ed.), *Computable Universe: Understanding and exploring Nature as computation*, World Scientific, Singapore, 2012.

[MAY 01] MAYNARD-SMITH J., *La Construction du vivant*, Cassini, Paris, 2001.

[NAG 12] NAGEL T., *Mind and Cosmos*, Oxford University Press, Oxford, 2012.

[NIE 08] NIETZSCHE F., *On the Genealogy of Morals*, Oxford University Press, Oxford, 2008.

[OCO 15] O'CONNOR T., WONG H.-Y., "Emergent properties", *Stanford Encyclopaedia of Philosophy*, 2015.

[ORU 14] ORUS R., "A practical introduction to tensor networks: matrix product states and projected entangled pair states", *Annals of Physics*, vol. 349, pp. 117–158, 2014.

[POP 73] POPPER K., *La logique de la découverte scientifique*, Payot, Paris, 1973.

[PRI 96] PRIGOGINE Y., *La fin des certitudes,* Odile Jacob, Paris, 1996.

[RIV 89] RIVAIL L., *Chimie quantique*, InterEditions, Paris, 1989.

[SPE 58] SPENGLER O., *L'homme et la technique*, Gallimard, Paris, 1958.

[SUS 15] SUSSKIND L., HRABOVSKY G., *Le minimum théorique*, Presses polytechniques universitaires romandes, Lausanne, 2015.

[SUS 16] SUSSKIND L., FRIEDMAN A., *Mécanique quantique*, Presses polytechniques universitaires romandes, Lausanne, 2016.

[TON 91] TONNELAT J., *Thermodynamique probabiliste*, Masson, Paris, 1991.

[TON 15] TONONI G., KOCH C., "Consciousness: here, there and everywhere?", *Philosophical Transactions of the Royal Society B*, vol. 370, 2015.

[WAL 16] WALKER S.I., KIM H., DAVIES P.C.W., "Informational architecture of the cell", *Philosophical Transactions of the Royal Society A*, vol. 374, 2016.

[WEI 11] WEISBERG M., NEEDHAM P., HENDRY R., "Philosophy of chemistry", *Stanford Encyclopaedia of Philosophy*, 2011.

[WEI 14] WEI W. *et al.*, "Study of exotic ions in superfluid helium and the possible fission of the electron wave function", *Journal of Low Temperature Physics*, vol. 178, pp. 78–117, 2014.

[ZEN 16] ZENG B., CHEN X., ZHOU D.-L. *et al.*, "Quantum information meets quantum matter – from quantum entanglement to topological phase in many-body systems", arXiv eprint 1508.02595, 2016.

[ZUR 14] ZUREK W.H., "Quantum Darwinism, classical reality, and the randomness of quantum jumps", *Physics Today*, vol. 67, pp. 44–50, 2014.

Index

A, B, C

D, E, F

G, H, I

K, L, M

N, O, P

Q, R, S

T, U, V

W, Z

Other titles from

ISTE

in

Science, Society and New Technologies

2018

BARTHES Angela, CHAMPOLLION Pierre, ALPE Yves
Evolutions of the Complex Relationship Between Education and Territories (Education Set - Volume 1)

GEORGANTOPOULOU Christina G., GEORGANTOPOULOS George A.
Fluid Mechanics in Channel, Pipe and Aerodynamic Design Geometries 1 (Engineering, Energy and Architecture Set – Volume 2)

GEORGANTOPOULOU Christina G., GEORGANTOPOULOS George A.
Fluid Mechanics in Channel, Pipe and Aerodynamic Design Geometries 2 (Engineering, Energy and Architecture Set – Volume 3)

2017

ANICHINI Giulia, CARRARO Flavia, GESLIN Philippe, GUILLE-ESCURET Georges
Technicity vs Scientificity – Complementarities and Rivalries (Social Interdisciplinarity Set – Volume 2)

DUGUÉ Bernard
Information and the World Stage – From Philosophy to Science, the World of Forms and Communications
(Engineering, Energy and Architecture Set – Volume 1)

GESLIN Philippe
Inside Anthropotechnology – User and Culture Centered Experience
(Social Interdisciplinarity Set – Volume 1)

GORIA Stéphane
Methods and Tools for Creative Competitive Intelligence

KEMBELLEC Gérald, BROUDOUS EVELYNE
Reading and Writing Knowledge in Scientific Communities– Digital Humanities and Knowledge Construction

MAESSCHALCK Marc
Reflexive Governance for Research and Innovative Knowledge
(Responsible Research and Innovation Set - Volume 6)

PARK Sejin, GUILLE-ESCURET Georges
Sociobiology vs Socioecology – Consequences of an Unraveling Debate
(Interdisciplinarity between Biological Sciences and Social Sciences Set - Volume 1)

PELLÉ Sophie
Business, Innovation and Responsibility
(Responsible Research and Innovation Set - Volume 7)

2016

BRONNER Gérald
Belief and Misbelief Asymmetry on the Internet

EL FALLAH SEGHROUCHNI Amal, ISHIKAWA Fuyuki, HÉRAULT Laurent, TOKUDA Hideyuki
Enablers for Smart Cities

GIANNI Robert
Responsibility and Freedom
(Responsible Research and Innovation Set - Volume 2)

GRUNWALD Armin
The Hermeneutic Side of Responsible Research and Innovation
(Responsible Research and Innovation Set - Volume 5)

LAGRANA Fernando
E-mail and Behavioral Changes – Uses and Misuses of Electronic Communications

LENOIR Virgil Cristian
Ethical Efficiency – Responsibility and Contingency
(Responsible Research and Innovation Set - Volume 1)

MAESSCHALCK Marc
Reflexive Governance for Research and Innovative Knowledge
(Responsible Research and Innovation Set - Volume 6)

PELLÉ Sophie, REBER Bernard
From Ethical Review to Responsible Research and Innovation
(Responsible Research and Innovation Set - Volume 3)

REBER Bernard
Precautionary Principle, Pluralism and Deliberation – Sciences and Ethics
(Responsible Research and Innovation Set - Volume 4)

VENTRE Daniel
Information Warfare – 2nd edition

Printed and bound by CPI Group (UK) Ltd, Croydon, CR0 4YY